HALKA CHRONIC

PAGES
OF
STONE

GEOLOGY OF WESTERN NATIONAL PARKS & MONUMENTS

3: THE DESERT SOUTHWEST

The Mountaineers • Seattle

THE MOUNTAINEERS: Organized 1906
*". . . to explore, study, and enjoy
the natural beauty of the outdoors."*

Published by The Mountaineers
306 Second Avenue West, Seattle, Washington 98119

Published simultaneously in Canada by Douglas & McIntyre
1615 Venables Street, Vancouver, British Columbia V5L 2H1

Cover design by Constance Bollen
Layout by Bridget Culligan
Cover photos: Dissected alluvial fan at Death Valley and
footprints at White Sands National Monument.
Photos by author unless otherwise credited
Manufactured in the United States of America

Library of Congress Cataloging in Publication Data
(Revised for Vol. 3)

Chronic, Halka
    Pages of Stone

    Includes bibliographies and indexes.
    Contents: 1. Rocky Mountains and Western Great
Plains   —   2. Sierra Nevada, Cascades, and Pacific
Coast   —   3. The desert Southwest.
    1. Geology—West (U.S.) 2. Geology—Great Plains.
3. National parks and reserves—West (U.S.).   4. National
parks and reserves—Great Plains.   5. Natural monuments
—West (U.S.)   6. Natural monuments—Great Plains.
I. Title.
QE77.C57 1984   557.3   82-422
ISBN 0-89886-095-4 (pbk.: v. 1)
ISBN 0-89886-114-4 (pbk.: v. 2)
ISBN 0-89886-124-1 (pbk.: v. 3)

0 9 8 7 6

5 4 3 2 1

Contents

To Tad and Mary Jane

Where wind meets sand, dunes climb skyward. These, composed entirely of gypsum grains, are in White Sands National Monument.

PREFACE

In the southwest deserts—a province that extends from southern Idaho to Mexico—the Earth's storybook stands open before us, each mountain range stark and sharply drawn, each arid valley displaying geologic features common to deserts everywhere or unique to deserts of the American Southwest. Geologic studies of the ranges and basins date back little more than a century, to times of prospectors and miners searching for nature's hidden treasures. After them came settlers, who of necessity sought out springs and likely sites for wells. From prospectors and settlers to scientists interested in knowledge for the sake of knowledge has come much of what we know about this country today. Thorough geologic mapping of the Southwest became the task of the United States Geological Survey, whose geologists hiked many a hot and weary mile to produce geologic maps and treatises available to all.

Now, long after the prospectors and early settlers, amidst thriving cities and fertile fields, geology is being looked at anew, this time by "laymen"—residents and travelers, non-professionals eager to learn about their surroundings: trees and birds and desert wildflowers, winds and clouds, and the very rocks and landforms of the ranges and the deserts between.

Spurred by geologic "happenings" like the eruption of Mount St. Helens in 1980, the disastrous Mexico City earthquake and Colombian volcanic eruption of 1985, and predictions of devastation to come along the San Andreas Fault in California, interest in the make-up of our planet has burgeoned. More and more of us—both geologist and non-professional—are looking now at the realities of our unstable Earth, feeling the impetus to understand its whims and vagaries. Recent advances in geologic thinking, including development of the Theory of Plate Tectonics with its drifting continents and widening oceans, and new hypotheses regarding evolution and worldwide extinctions, constantly filter, nowadays, through the media, to appear in popular magazines and on TV screens throughout the nation. Geology is here. Geology challenges. Its challenge can no longer be denied.

But many of the stories told by mountains and valleys are for the average observer far from clear, even where rocks are well exposed. So this volume, like its predecessors in the *Pages of Stone* series, seeks to lead its readers in a search for greater understanding of the Earth. And in unraveling the Earth's story, it draws its contents from geologic literature, from discussions with geologists familiar with the southwestern Basin and Range region, and from personal observations by the author. It draws its examples from our western national parks and monuments, where scenic beauty is born of, and reflects, the geologic story.

These park areas, administered by the National Park Service, are preserved as closely as possible in their natural state: Animals and plants, rocks and rivers, cliffs and chasms, are virtually pristine, as they were when Europeans first saw and admired them. Thus the parks and monuments are proper places to begin a study of the Earth in its many forms and fancies. As the national park system developed, it was early recognized that parks and monuments should include more than just scenery: They should embrace areas representing a variety of natural features—both geologic and biologic—as well as historic features, and they should serve not only as "pleasuring grounds" but also as outdoor lecture halls and laboratories where visitors may gain on-site knowledge of the story of our planet and of the plants and animals that share it with us.

In the park areas—both parks and monuments—are illustrated most of the principles on which the study of geology is based. In them are rocks formed by long, slow cooling of molten magma, and volcanoes new and old, active, dead, or dangerously dormant. In them are illustrations of sedimentary processes by which new rock is made from old. In them are rocks altered by high temperatures and great pressures, examples of changes going on even today, invisibly, deep in the crust of the Earth. In them are also many examples of faults and folds and other geologic structures created during the endless battle between forces of uplift and the process of erosion. In them are landforms devolving from weathering and erosion of "hard" rock and "soft" rock, flat rock and tilted.

Geology is a logical science, a science that studies the shapes and compositions of rocks and their relations to one another, to put together, one by one, the torn and far-flung pages of the Earth's great history book. Like all sciences, geology constantly builds new data, new ideas, on earlier findings. Geologists search for clues to the past and add them to a long line of clues that has come before. They seek to determine the nature of the rocks, and to interpret it in terms of the rocks' origins. With relatively new-found tools they work out the ages of rock samples, which directly relate to their probable origins, as well as to the history of the continents and sea floors, the history of the planet as a whole. Many geologists, working in many areas, hiking, driving, flying, even diving, gathering quantities of data, writing many scientific papers and treatises, discussing their data and their interpretations with colleagues, are slowly filling in the story of our island planet and of its wondrous product, life.

Geology unlocks the history of the Earth by using the present as the key to the past. As portions of the Earth's crust rise and fall today, or break along great belts of faults, or push under or

over other portions (as has happened so recently in Mexico), the Earth shudders and continents move and mountains thrust upward—as they have done over many millions, even billions, of years. Today's changes may be barely discernible, too small to measure, but often-repeated changes over long periods of time have broken continents apart, submerged them, lifted them high, crashed them into one another, widened or narrowed seas, and built high mountain ranges and deep valleys. As mountains rise, erosion tears at their flanks, today as yesterday, yesterday as a million years ago. Wind and water and frost hammer at the heights, breaking down the rocks, washing the fragments downstream. Rain becomes rivulets, rivulets become streams, streams become rivers, carrying the materials of the mountains to the sea or into low interior basins. Rocks and sand and silt carried by the rivers are deposited on floodplains, lake floors, and deltas far from their source, today as in the past, forming layered strata that will harden into sandstone, shale, and limestone. And in these strata, present and past, is buried a record of life—the shells and body parts of animals and plants, parts of our geologic puzzle, evidence of the evolution of life.

A few special comments: In this book, in accordance with National Park Service usage, all measurements are given in metric units, with English equivalents in parentheses. Geologic terms are defined where first used, as well as in the glossary at the end of the book. In Part 1, where many of the most common geologic concepts are introduced, new terms are printed in **bold lettering**.

Park and monument regulations prohibit collecting of plants, animals, fossils, or just plain rock. They also ask you to carry out what you brought in, especially in the way of lunch sacks, paper napkins, cigarette butts, and other unsightly trash. Please comply with these regulations. An observant eye, an inquiring mind, sturdy shoes, perhaps a camera or a pair of binoculars, are good companions in the desert parks and monuments.

As you explore park areas, do allow time to stop and examine the rocks. Look, too, at landforms—mountains, mesas, peaks, plains, and valleys—and ponder their origins. I encourage you to get out of your car, to walk and wander, for more leisurely looks at both rocks and scenery. There are enticing trails in almost every park and monument, many with trailside exhibits or with brochures describing natural features along the way. When walking on the desert, remember that among the rocks, among the many interesting desert denizens, there are just a few venomous reptiles and insects. Watch where you put your hands and feet—a practice that will also save you from innumerable cactus spines. If you plan to hike any distance from established roads and trails, tell someone where you are going, take frequent bearings on prominent landmarks, and carry water: It's mighty thirsty out there.

I sincerely thank the many friends, colleagues, and National Park Service personnel who have helped me with this book, in particular Lehi Hintze and Keith Rigby of Brigham Young University, Wesley Peirce of the Arizona Bureau of Mines and Mineral Technology, Bennie Troxel of the University of California, Frank Kottlowski and his colleagues at the New Mexico Bureau of Mines and Mineral Resources, Robert B. Raup of the U.S. Geological Survey, and Andrea Eddy, all of whom waded through part or all of the manuscript, adding many comments and suggestions.

My gratitude also extends to Tad Nichols, who freely contributed both photographs and photographic advice; to my daughter Emily Silver, responsible for many of the drawings and diagrams; and to the friends and relations whose hospitality and companionship cheered the work along.

Safety is an important concern in all outdoor activities. No guidebook can alert you to every hazard or anticipate the limitations of every reader. Therefore, the descriptions of roads, trails, and natural features in this book are not representations that a particular place or excursion will be safe for your party. When you visit any of the places described in this book, you assume responsibility for your own safety. Under normal conditions, such excursions require the usual attention to traffic, road and trail conditions, weather, terrain, the capabilities of your party, and other factors. Keeping informed on current conditions and exercising common sense are the keys to a safe, enjoyable outing.

The Mountaineers

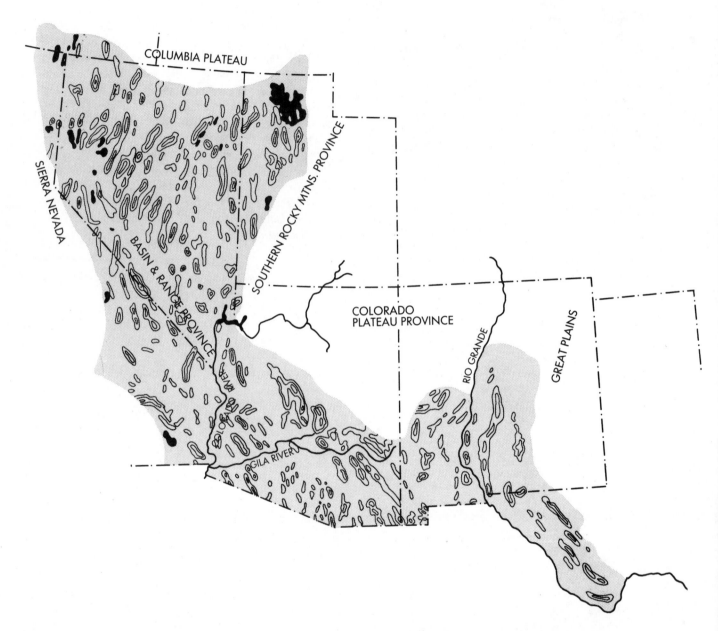

The ranges of the Basin and Range Province swing south and southeast from Oregon to Mexico.

PART 1.

GEOLOGY FOR EVERYONE

I. The Face of the Desert

The vast deserts of the Southwest—the arid basins and gaunt ranges—have an allure all their own. Sunny skies and mild winters attract ranchers in search of natural forage for their herds; farmers pair winter sunshine with irrigation water to produce year-round crops. They also attract "snowbirds" by the thousands, human migrants seeking respite from northern winters. Yet many valleys are empty still, practically untouched by Man, and many mountain ranges are devoid of established roads or trails.

The region known to geologists and geographers as the Basin and Range Province reaches from southern Oregon to the Big Bend of the Rio Grande in Texas. It embraces parts of California, Nevada, Idaho, Utah, Arizona, New Mexico, and Texas. Its name is descriptive, for here broad valleys alternate with steep, rugged ranges, some

In the Basin and Range Province, inadequate rainfall reduces the vegetation to specially adapted plants such as cacti and almost leafless shrubs.

high enough to support pine and aspen forests near their summits, others mere ridges interrupting the desert floor. Some few of the valleys—notably those of the Colorado River and the Rio Grande—are watered year-round by major rivers; in others, streams flow only during and after heavy rains, or with the melting of highland snow.

Many valleys are completely ringed by mountains and have no outlets, no way to drain. Streams from surrounding ranges sink into porous basin gravels or create shallow, temporary **playa lakes** at their lowest point. In many of these enclosed valleys, lake sediments and abandoned shorelines show that larger lakes existed in the past.

Just what *is* a desert? One definition—and a good one from a biologist's point of view—describes a desert as an area in which too little and too sporadic rainfall has left a strong impression on plant and animal life. To this we must add that shortage and irregularity of rainfall have also strongly affected geologic processes such as weathering, erosion, and the development of landforms.

Characteristically, deserts develop where evaporation exceeds precipitation, often several times over. Even along the few permanent streams, all of this land is short of water, deprived of it by a relatively dry high-pressure cell in the atmosphere over the eastern Pacific, and by the great barrier of California's Sierra Nevada—in places more than 3 kilometers (2 miles) high—and ranges south of it. Intercepting what moisture does come from the west, this mountain barrier forces the air to rise, become cooler, and shed its moisture in the high country. As the winds descend the east slope of the Sierra, they warm up again, and could again hold more water vapor. They thirstily absorb what they can of the desert's slim supply of moisture, intensifying the perennial drought. Only during two parts of the year does the desert receive significant amounts of rain: in winter, when the east Pacific high-pressure cell follows the sun southward; and in summer, when moist winds sweep in from the south, where there are no major mountain barriers to steal their moisture.

Despite the low and sporadic rainfall, many plant and animal species do live out their lives on the desert. Through the millions of years that these deserts have existed, several groups of plants have adapted to the arid climate by deepening their root systems and reducing, in one way or another, their evaporative surfaces. Some, like ironwood and mesquite trees, sprout only the tiniest of leaves. Others, like the creosote bush, veneer equally tiny leaves with a water-

The Colorado River, here seen just below Parker Dam, brings Rocky Mountain water to a thirsting land.

Racetrack Playa, in Death Valley National Monument, gleams white in the desert sun. Inset shows the playa's dry, mudcracked surface.

Hardy trees such as palo verde, mesquite, and ironwood find nourishment beneath the usually dry sands of desert washes. This stream flows only after heavy rains.

proof layer of waxy resin. Ocotillos, scrawny clusters of spiny sticks, sprout their leaves after good soaking rains; the remainder of the year their emaciated stalks appear dry and dead. Cacti have given up leaves altogether; their thick, green, knobby or accordion-pleated stems, which expand to store water, are furnished with the chlorophyll needed for photosynthesis. And many small annuals produce seeds that lie dormant for years, awaiting a wet spring to sprout and bloom and seed again. On rocky mountain slopes many plants grow almost without soil, their roots penetrating rock-bound crevices in search of moisture and mineral nourishment. Higher mountaintops, receiving more rain and snow, support substantial forests on better developed soils.

Animals too have adjusted to the desert's ways. Many hide during the day in burrows or under shrubs and trees, then emerge at night to forage, mate, and play. Others limit their activities to dry but shaded stream beds (called "washes" in the Southwest) or to cool dawn and evening hours. Yet others sleep the summer through in cool underground burrows.

Geologically this arid land is different, too. Mountains are barren and skeletal, without the smooth flesh of soil and vegetation. Their rocks are clearly exposed, though commonly darkened with **desert varnish**, a thin, hard, often shiny, dark brown or blue-black polish of iron and man-

ganese oxides. Desert varnish develops slowly, over many centuries, either from minerals gradually leached from the rock itself or from soluble material carried in desert dust and spread thinly, over and over again, by occasional rain. Many desert surfaces are armored with **desert pavement**, a thin layer of pebbles left behind when all finer material has been blown away by the wind.

Because rain and snow are relatively rare, and because soils as a result are thin and vegetation sparse, desert landforms look different from those found in moister climates. Most streams flow, as we have seen, only during and immediately after heavy rains. Because **weathering**, the slow decomposition and disintegration of rock, creates more rock debris than the intermittent streams can carry, mountain streams when they *do* flow are overloaded with boulders, gravel, and sand. Using this rock debris for tools, the rushing waters gouge steep, angular canyons. But at the mouths of the canyons, where the gradient suddenly lessens, the streams slow down abruptly, drop their loads, and sink into the coarse gravel deposited by previous storms (thereby recharging groundwater levels of the basins). The rock material carried by these streams gradually builds up as **alluvial fans** at the mouths of canyons.

As alluvial fans build up and out, they merge with those from adjacent canyons to form broad, gentle, gravel-covered aprons known in the

Finely broken white quartzite makes up a desert pavement that armors the desert against further wind erosion.

Though this stream flows only after a rain, vegetation flourishes along its banks, helping to control erosion.

The surface of an alluvial fan is strewn with boulders, cobbles, and pebbles transported from nearby mountains.

Southwest as **bajadas** (ba-HA-duhs)—a name borrowed by geologists from the Spanish who settled this region.

Much desert erosion occurs between storms, without the work of running water. Weathering processes gradually break down the rocks, acting on exposed surfaces and along fractures inherent in almost all rock masses. Angular blocks may then be pried loose by winter frosts, by swelling of rain-wet clay, by the roots of wind-swayed plants, or by the burrowing of animals. The blocks bounce down mountain cliffs, often initiating **rockslides**. Pebble- and sand-sized particles also slip and slide. Slopes not held in place by plant roots (and so many are not!) may give way in **slumps** or **landslides**. Gradually, through the centuries, these processes reduce the size of the desert ranges. As mountains wear down, valleys are more and more deeply filled with rock debris from the mountains.

Wind plays a strong role in desert erosion and the creation of desert landforms. Even gentle winds can pick up clay and silt from the desert floor, and lift them to great heights. Strong winds of winter and spring can lift sand particles as well, bouncing them along close to the ground, using them to sandblast other particles from exposed rock surfaces. Coarse sand grains and even small pebbles are moved by the strongest winds, such as those that spiral upward in summer "dust devils"; such winds and their airborne gravel can pit a rock surface (or a car windshield) in no time flat, as anyone who has been caught in one knows. Wind abrasion repeated over time may produce deep hollows and grooves in exposed rock surfaces, undermine

Trees growing in joints gradually force rock apart. Here the roots penetrate two sets of parallel joints, one set nearly horizontal, the other vertical.

Angular blocks of basalt move slowly downhill in rock streams. At the bottom of the photograph, floodplain sediments have been sliced away by the same river that deposited them.

Near White Sands National Monument, strong spring winds swirl fine gypsum dust into the air.

Desert landforms include flat-topped mesas and gently sloping cuestas capped with resistant rock layers.

cliffs, or help to form strange mushroom-shaped **balanced rocks**.

Many of the basins of the Southwest started out as undrained depressions. As they filled with rock debris, their streams and rivers joined with those of other basins, so that through drainage was established. With through drainage, streams cut down through their own older deposits, with slow destruction of many alluvial fans and bajadas. Some basins, however, remain undrained to this day.

Among the most fascinating features of erosion are those formed underground by solution of limestone. **Groundwater** seeping through tiny cracks and crevices gradually dissolves away part of the limestone, slowly enlarging its crevices and shaping passages and caverns. Later, if the water level drops and the caverns become air-filled, rain water trickling down from the surface may deposit intricate **flowstone** and **dripstone** ornaments on cavern surfaces. Most desert caverns date back to times when both rainfall and groundwater levels were higher than they are today.

II. Basins and Ranges

What of the larger landforms of the desert—the basins and ranges themselves? How did they come into being? Many of the mountains are, in geologic parlance, **fault-block ranges**. **Faults** are breaks in the Earth's crust or in rocks of the crust, *along which movement has taken place*. Here in the desert southwest various segments of the crust broke apart along faults, then tipped and tilted like children's blocks. Some segments pushed upward into mountains, some dropped downward to form the basins between, and some traveled long distances more or less horizontally.

Faults come in many shapes and sizes. Movement on large ones may build or displace mountains; fault movement can also create coastlines, delineate large plateau areas, or help establish river courses. Small faults may represent movements as small as a millimeter (about 1/16 inch).

Geologists recognize several basic kinds of faults, as shown on the accompanying diagram. Most types can be found in the southwest deserts. Blocks of crust that are thrust upward between pairs of normal or reverse faults are called **horsts**; portions dropped downward between normal or reverse faults are **grabens**. Most of the desert fault-block ranges are horsts, and most of the basins between them grabens. Normal faults are thought of as pull-apart features, as are low-angle (nearly horizontal) extensional faults. These two types occur where there is tension in the crust as a whole. Extensional faults in the

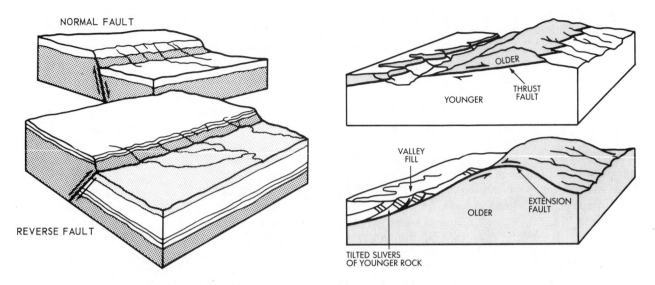

Faults of many kinds give character to the Basin and Range Province. Several cycles of faulting are well represented in national parks and monuments within this region.

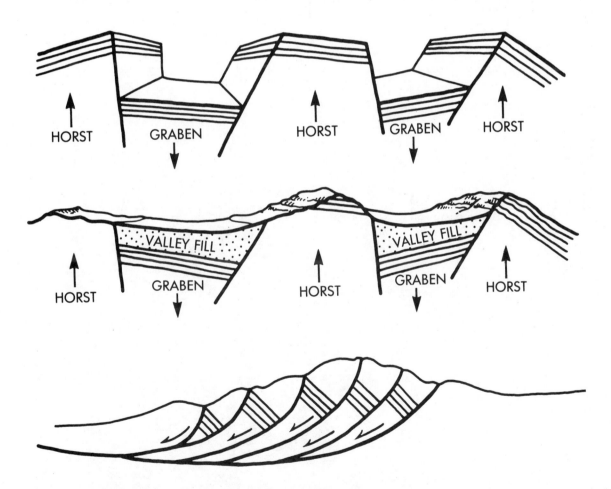

To the first geologists who worked in the Southwest, the basins and ranges were grabens and horsts. Recent discovery of a seismic reflector (a change in rock density that reflects man-made shock waves) suggests that faults are curved and flatten out with depth. Moving along curving slide planes, once-horizontal rocks become tilted.

western third of North America may have widened the continent by as much as 300 to 500 kilometers (200 to 300 miles). Reverse faults and low-angle thrust faults are push-together structures and occur where the crust is compressed horizontally. Strike-slip faults mark sideways movement such as that shown along the San Andreas Fault in California. The upward or downward or sideward movement along faults is relative: One side may really have moved in the opposite direction from the other, or one of the sides may merely have moved farther than the other in the same direction. And faults are not necessarily flat and straight; curved ones, which tend to tilt overlying rocks (see figure), are responsible for many of the Southwest's tilted ranges.

Most of the fault-block ranges of the Southwest are longer than they are wide, and most of them lie with their long axis trending north-south or northwest-southeast or (in Nevada) northeast-southwest. The actual faults that define these ranges may be far out from the present margins of the mountains, as erosion has cut back into the mountain mass. Not all the mountains of the Southwest are fault-block ranges. Some are volcanic, created from molten outpourings from the Earth's interior. Volcanism appears in special concentration along two **rift valleys**, the Rio Grande Rift and the Salton Sea Graben. Both of these long, slender, very deep fault valleys bear special meaning for some of the national monuments discussed in this book.

The gaunt desert ranges are a geologist's joy. The rocks stand out as barren cliffs and crags, marked by time, ready to be seen and sampled and studied. But—and it is a big but—the craggy ranges are disconnected, separated by broad sediment-filled valleys that effectively conceal, with thousands of meters of sand and gravel, silt and salt, the buried connections between the ranges. The desert is a giant patchwork quilt with most of the pieces concealed from our view. What pieces do remain at the surface reveal a

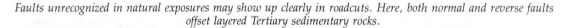

Faults unrecognized in natural exposures may show up clearly in roadcuts. Here, both normal and reverse faults offset layered Tertiary sedimentary rocks.

complex history of successive invasions of the sea, successive episodes of mountain-building, and successive periods of erosion.

Throughout these deserts one can discern certain set geologic patterns. Maps, topographic or geologic, reveal the parallel orientation of most desert ranges. In parts of this vast region, groups of ranges all display the same types of rock, or show similarities in the arrangements of their rocks, their geologic **structure**. Some are broad arches of once horizontal rock layers deposited on the floors of ancient seas. Some are only parts of arches, and seem once to have been conjoined with companion ranges across intervening valleys. Some are great dome-shaped masses of once molten granite, with thin shells of crushed and altered rock, in a pattern now known to geologists as **metamorphic core complexes**. Some are volcanoes, others the mere remnants of volcanoes. Yet others repeat the flat-topped profiles caused by hard, relatively young lava-flow caps. And among all the aligned ranges are some completely unaligned ones composed almost entirely of solidified volcanic ash.

Let's take a moment here to discuss another trio of structures seen in the rocks: various types of **folds**. It seems hard to believe that rocks can bend and fold, but there is ample evidence that they can. Great pressure, applied over a long-drawn-out period of time, slowly bends even the staunchest rock. Heating—such as that which oc-

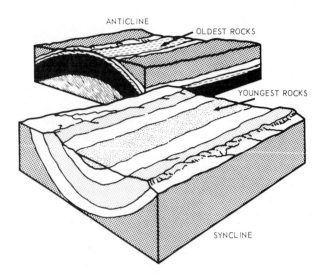

curs deep down in the Earth's crust—helps, just as heating a steel bar in a forge makes bending it into a horseshoe possible. And just as geologists identify several types of faults, they recognize several types of folds, shown on the accompanying diagram. These types may occur alone or in combination. **Monoclines** may be large or small. (They come into their own in the Grand Canyon-Colorado Plateau country, to be described in Volume 4 of this series.) Both **anticlines** and **synclines** may be quite broad, or may be squeezed into tight accordion pleats. There are good ex-

Until recently, geologists have been hard put to explain the great granite domes of metamorphic core complex ranges.

Once flat-lying, these layered rocks are bent into a fold by mountain-building forces.

amples of all of these types in the Southwest's desert ranges.

Clearly, geologic patterns give us our scenery here. Sometime in the past, the mountains have risen, tilted and jostled like children's toys. The forces of weathering and erosion—water, wind, and frost—have shaped their cliffs and spires, worn their ravines and canyons, etched and sharpened hard rocks, washed softer, less resistant ones away.

Before we go farther we need to look at the Earth as a whole, to see what is going on beneath its crust that can influence developments on its surface.

III. Core, Mantle, and Crust

What caused this great jumble, this desert patchwork quilt? Hidden from view, the interior of the Earth long remained a geologic mystery. However, geologists have now found clues to its makeup in earthquake waves that pass through its interior, and in its total mass as determined by its gravitational effect on the moon and on other planets.

What have we learned from these clues? We've learned that the Earth's interior is cored with a sphere of almost pure nickel and iron, its center solid but its outer part semi-liquid, capable of some degree of plastic movement, as is, for instance, the malleable red-hot iron on a black-smith's anvil, or a cube of butter that settles and spreads, without quite melting, on a hot summer day.

Outside the core is another thick layer, the **mantle**. It is composed of **basalt**, the same type of black or dark gray rock that spills out on the Earth's surface from time to time and from place to place as dark gray or black **lava**. As is the case in the core, the inner part of the mantle is solid, the outer part liquid or semiliquid, a seething, half-molten, slow-motion boiling kettle of jam. The outer part of the mantle is the Earth's "stove," heated by decay of radioactive elements concentrated there. Convection currents develop in the upper mantle in the form of huge cells that surge upward in giant boils, roll over, and plunge downward again.

Powered by its own hot stove, the Earth's churning mantle strongly influences, as we shall see, the outermost shell of the Earth, the **crust**. The crust is made up of what we think of as solid rock. It seems to divide, on a worldwide scale, into two major types: The first type, **oceanic crust**, is dark gray or black and heavy with iron and magnesium, very like the mantle. It occurs, as its name suggests, beneath ocean basins. Oceanic crust is quite thin—something on the order of 5 kilometers (a mere 3 miles).

The other type of crust is **continental crust**. Lighter than oceanic crust in both color and weight, it is composed of minerals low in magnesium and iron. It occurs, as you have probably guessed, on continents, extending out to the limits of the continental shelf. It also occurs on microcontinents, big oceanic islands like New Guinea, Cuba, and Madagascar. Continental crust is thick—on the order of 30 to 40 kilometers (20 to 25 miles).

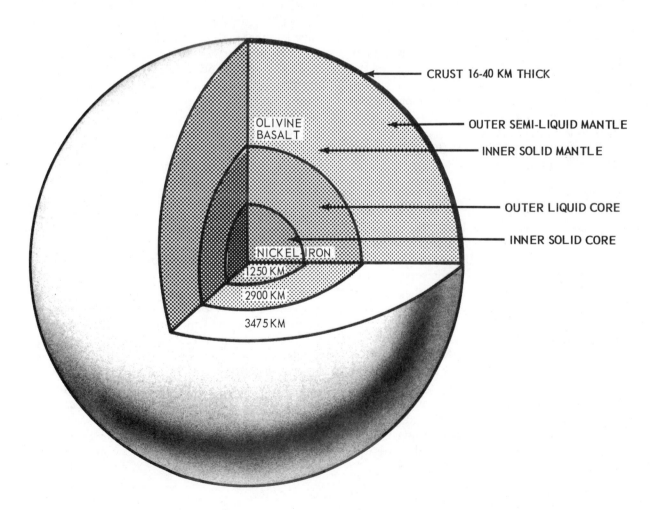

IV. A Theory That Works

The idea of **continental drift** was born early in this century—a logical idea stemming from recognition of the match in profiles of the continents bordering the southern Atlantic Ocean. Africa and South America, in particular, look as though they are two adjacent pieces of one jigsaw puzzle. And if you juggle other pieces a little bit, and include their continental shelves, Europe and North America can be fitted together, too. Some researchers thought they had once been connected, but had broken and drifted apart. Yet for many years no one could come up with a plausible theory that would allow portions of the thin, fragile crust, with no power plants of their own, to sail off around the globe like so many *Queen Marys*.

In the last two decades, however, geologists have developed some new views, born of the old. The **Theory of Plate Tectonics**, which took the geologic world more or less by storm in the 1960s and is now accepted by geologists everywhere, once again sets the continents adrift. It starts out by stating that the crust is coupled with the uppermost part of the mantle to form a stiff layer, the **lithosphere** ("rock sphere"), a layer about 60 kilometers (40 miles) thick under the ocean and about 90 kilometers (60 miles) thick under the continents. Relative to the size of the Earth, the lithosphere is very thin, just a film. Like the film that forms on a bowl of chicken broth, it can be rumpled and moved about by stirrings in the liquid below.

The theory also states that the lithosphere, and therefore the crust, is made up of a dozen large **plates**, fitted together like the plates of a turtle's shell, with a bevy of smaller plates squeezed in here and there among the large ones. The plates are separated along suture lines that in some cases are submerged volcanic mountain chains of **mid-ocean ridges**, and in other cases are either deep-sea **trenches** like those off the west coast of the Americas, or world-class mountain ranges like the Himalayas or the Alp-Carpathian-Caucasus chain. It goes along with the theory that all these regions are beset with frequent earthquakes and periodic volcanic activity.

Geologists have learned as well that the Earth's crust is not all as ancient and permanent as they used to think. New crust is constantly coming into being in the middle of the sea, along mid-ocean ridges that thread their way down the Atlantic, through the Indian Ocean, and across the Pacific. On the ridges, which can be likened to long lines of volcanic fissures, **magma** (molten rock) boils up from the mantle and hardens along the suture between two crustal plates. The magma is able to push up this way because the crustal plates on each side of the mid-ocean ridge move apart, apparently dragged along by powerful but slow-moving currents stirring the half-molten upper part of the mantle. Created in narrow bands, the new crust is then pushed apart by still newer lava. Continuous replenishment makes of the oceanic crust two broad conveyor belts moving slowly in opposite directions away from the mid-ocean ridge. This process, known as **sea-floor spreading**, is a vital part of the Theory of Plate Tectonics.

The Earth is not getting any larger. So if new crust is born along mid-ocean ridges, to compensate, old crust must somewhere be destroyed. If continents drift apart, plates must somewhere collide. Where one slowly moving continental plate bumps into another, mountains are pushed up, foreshortening the crust. This has happened, for instance, where India, once a small continent in itself, bumped into Asia: The Himalayas are the glorious result. Where an oceanic plate collides with a continental plate, the picture is a little different. The oceanic plate, being made of heavier material, is overridden by the lighter (though thicker) continental plate, and is then drawn down—**subducted**—to depths where it remelts and once more becomes part of the mantle. Some plates seem to pivot, or rotate, turning either clockwise or counterclockwise, scraping along the sides of other more stable plates, creating **strike-slip faults** with largely horizontal movement; the San Andreas Fault of California is an example.

Volcanism is likely to occur near the line where **subduction** takes place, with magma rising from two sources: the remelted oceanic crust, which gives rise to dark, iron-rich magma; and portions of the leading edge of the continental plate, drawn down and remelted along with the oceanic crust. The more buoyant continental material, once melted, may push upward through the crust at the edge of the continent. If such magma rises only part of the way to the surface, and cools while it is still at considerable depth, it hardens—crystallizes—into great masses of granite like those in the Sierra Nevada. Where magma breaks through to the surface, volcanoes form, with light-colored lava unlike the dark lavas derived directly from the mantle. Since the light-colored lavas are thick and sticky and more explosive than their dark-colored counterparts,

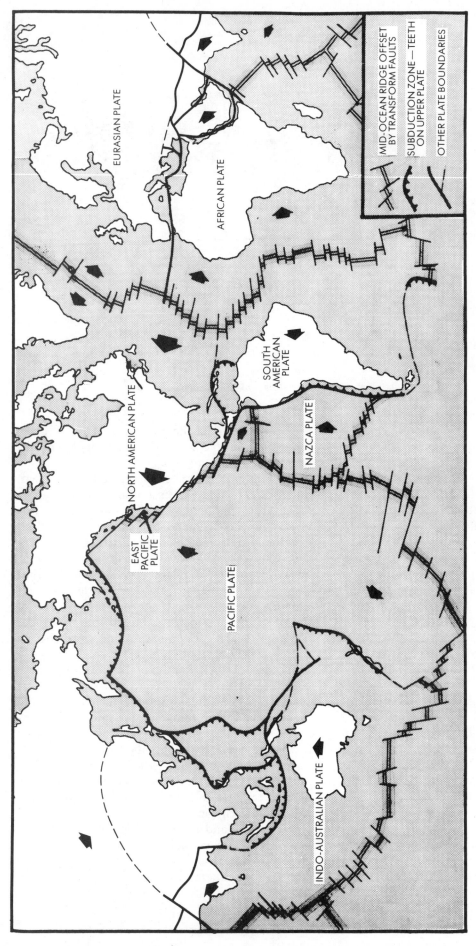

The Earth's crust is divided into a mosaic of large and small plates bounded by mid-ocean ridges and zones where collision is taking place. Plate margins are sites of frequent earthquakes and volcanic activity. Magnetic soundings show that mid-ocean ridges are offset by numerous transform faults.

Legend:

MID-OCEAN RIDGE OFFSET BY TRANSFORM FAULTS

SUBDUCTION ZONE — TEETH ON UPPER PLATE

OTHER PLATE BOUNDARIES

Plate labels:

EURASIAN PLATE

AFRICAN PLATE

NORTH AMERICAN PLATE

SOUTH AMERICAN PLATE

NAZCA PLATE

EAST PACIFIC PLATE

PACIFIC PLATE

INDO-AUSTRALIAN PLATE

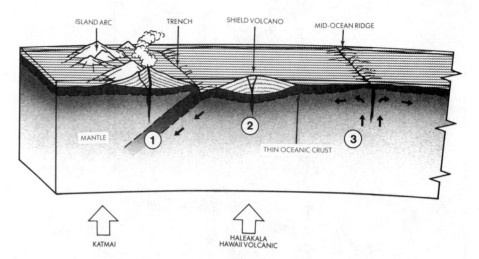

1. *Carried along by convection currents in the mantle, oceanic crust is drawn under and melted in subduction zones, forming basalt magma which erupts along island arcs.*

2. *Shield volcanoes form as plumes of basalt magma rise above isolated "hot spots" below the ocean floor*

3. *At mid-ocean ridges, basalt magma rises to form new oceanic crust.*

they pile up as tall volcanic cones that may burst forth again and again with violent explosive eruptions.

Many of the larger plates are partly oceanic, partly continental. True to this format, the North American Plate, which extends from the Mid-Atlantic Ridge to the Pacific coast, is oceanic in its eastern part, continental in its western part. For the last 60 or 70 million years, new crust has been forming along the Mid-Atlantic Ridge. As new crust forms, the Atlantic widens and the North American continent, once firmly attached to Europe and Africa, drifts westward and south-westward. The Atlantic Ocean is now about 5400 kilometers (3300 miles) wide as measured from Cape Hatteras to North Africa. Half of this distance, 2700 kilometers (1650 miles), represents the westward and southwestward movement of the North American Plate. The other half represents the movement of the Eurasian Plate. With a little arithmetic we find that North America has for 60 to 70 million years sailed into the sunset at an average rate of 4 centimeters (less than 2 inches) per year. It is still sailing.

Returning to the southwest deserts, we can now better visualize what made them what they are. The westward drift of the North American Plate put it on a collision course with the East Pacific Plate, with the plate margins caught in the crush. Because the North American Plate is composed of continental crust, lighter in weight than the oceanic crust of the East Pacific Plate, it overrode the East Pacific Plate, forcing it downward to remelt in the mantle.

The amount of material pushed (or drawn) downward in this way is nothing short of enormous. The continent, whose western shore was at one time well inland from the present coast, eventually transgressed so far across the East Pacific Plate that it overrode half of the Pacific Basin, even covering large portions of the East Pacific Rise, formerly a mid-ocean ridge (see figure). As this happened, the advancing edge of the continent apparently bumped into a number of large islands or microcontinents made of continental-type crust.

Because of their continental-type crust, these islands or microcontinents were too light to be subducted. So the continent just hung onto them, tacked them to its margin, and thereby enlarged its total area. Geologists call this process **accretion**, and it had a significant part to play in developing the complex geology of the western-most Basin and Range area, where several of the desert ranges include rocks that originally formed as Pacific islands.

The ranges and basins themselves, buckled by older earth movements, seem to have been created by pulling-apart processes—tensions—resulting from general uplift of the western part of the continent, uplift seemingly due to heat cells in the mantle. Most of the movement that directly created the ranges may have been not uplift of

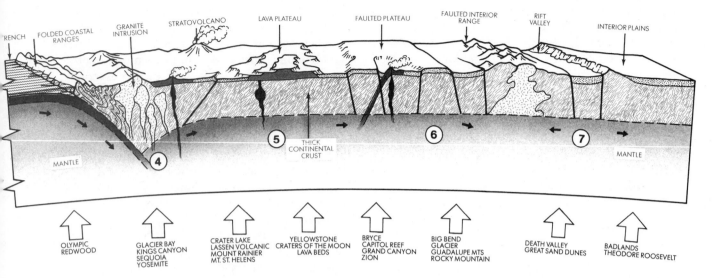

4. *Melting of continental crust along a subduction zone creates granitic magma, which may cool slowly, below the surface, in batholiths that are later bared by erosion. or the magma may erupt explosively to form stratovolcanoes.*

5. *Flood basalts, exceptionally fluid in nature, rise above "hot spots" below continental crust.*

6. *Interior ranges push upward in response to compression along the distant continental margin.*

7. *Rift valleys are tensional features where the crust drops as neighboring areas are pulled apart.*

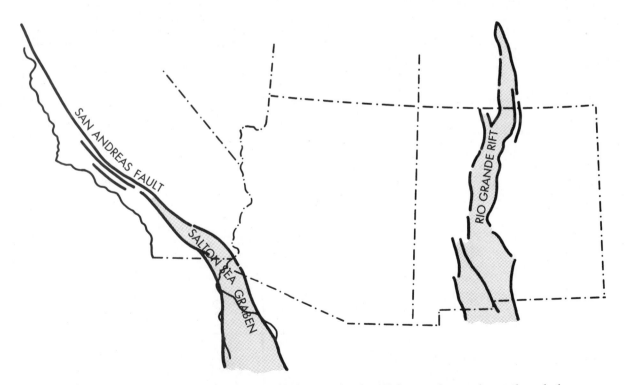

Two great rift valleys, bordered by major fault zones that extend down to the mantle, cut through the deserts of the southwest. The Salton Sea Graben extends south to form the Gulf of California; the Rio Grande Rift dictates the position of the Rio Grande.

the mountains but downdropping of the valleys. Two major down-faulted blocks, the long slivers of the Rio Grande Rift in New Mexico and the Salton Sea Graben in California, continuous with the Gulf of California, may be the bare beginnings of breaks that will eventually widen into new ocean basins. Their bounding faults are particularly deep, reaching all the way through the crust to the mantle. So we may be seeing, in their downdropped floors and in associated earthquakes and volcanoes, the birth of two new mid-ocean ridges, and of two narrow seaways that may someday evolve into new oceans.

V. Rocks and Minerals

Geologists recognize three main classes of rocks, all of them well represented in the ranges of the Southwest:

• **Igneous rocks** originate from molten rock material that rises from as much as 300 kilometers (200 miles) below the surface. Igneous rocks are further divided into two groups: those that cool and harden very slowly, deep below the surface, called **intrusive igneous rocks**, and those that harden more rapidly at the surface, called **extrusive igneous rocks** or, more simply, **volcanic rocks**. Long, slow cooling promotes crystal growth, so intrusive igneous rocks are generally coarsely grainy, with easily distinguished, interlocking crystals. Granite and its cousins granodiorite and quartz monzonite fall into this category. (It is perfectly legal to call all these coarsely crystalline rocks "granite," a practice adhered to in this book. The name shares its origin with "grainy" or "granular.") Volcanic rocks are generally very finely crystalline, with crystals that can't be discerned without magnification. Some volcanic rocks are glassy, with no crystals at all. Volcanic rocks include both **lava flows** and **tuff** (consolidated volcanic ash). As the chart shows, each intrusive rock has a matching, finer-grained volcanic counterpart.

• **Sedimentary rocks** form from the broken fragments or dissolved minerals of other rocks, transported and deposited by water, wind, or ice. Nearly always, sedimentary rocks are layered, or **stratified**, which makes them easy to

COMMON IGNEOUS ROCKS

INTRUSIVE IGNEOUS ROCKS				VOLCANIC EQUIVALENT		
NAME	COLOR	COMP.	MINERAL MAKE-UP	NAME	FLUID.	OCCURRENCE
GRANITE	light gray to white	increasingly silicic ⬆	mostly quartz and feldspars, with minor mica and/or hornblende	RHYOLITE	increasingly fluid when molten ⬇	fine, pale ash and tuff
QUARTZ MONZONITE, GRANODIORITE	light gray		mostly feldspars and quartz, with mica and hornblende	RHYODACITE		light gray ash or tuff, some volcanic domes
QUARTZ DIORITE	medium gray		mostly feldspars, with quartz, mica, and hornblende	DACITE		gray lava of volcanic domes, ash or tuff, pumice
DIORITE	medium gray		feldspars and quartz with abundant dark minerals	ANDESITE		short, thick lava flows, volcanic ash or tuff
DIABASE	dark gray to black		mostly dark minerals such as pyroxene, some plagioclase and quartz	BASALT		fluid black and dark gray lava flows, cinders

(GRANITE bracket spans GRANITE, QUARTZ MONZONITE/GRANODIORITE, QUARTZ DIORITE rows)

Moderately fluid lava forms shield volcanoes like this one near Capulin Mountain National Monument.

COMMON SEDIMENTARY ROCKS

NAME	DESCRIPTION
MUDSTONE	grains of silt and clay cemented together
SILTSTONE	grains of silt cemented together
SHALE	siltstone or mudstone that splits into flat slabs parallel to original bedding
SANDSTONE	grains of sand (usually quartz) cemented together
CONGLOMERATE	sand, pebbles, and cobbles deposited as gravel and later cemented together
LIMESTONE	calcium carbonate (calcite) rock deposited as limy mud or fragments of shells

spot in the hills and ranges. (Lava flows and falls of volcanic ash may be stratified, too.) Sandstone and siltstone deposited by flowing water or wind may be **cross-bedded**, with fine slanting laminae at an angle to the stratification. Sedimentary rocks frequently contain **fossils**, remains or traces of plants and animals that lived and died when the rock was formed. These rocks may also display ripple marks and mud cracks that formed on their surfaces at the time they were deposited.

Sedimentary rocks are classified by grain size and composition, as in the table. They can also be classed as **marine** if they are deposited in the sea, and **continental** if they are deposited by rivers, lakes, glaciers, or wind.

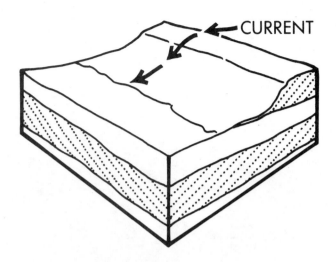

Cross-bedding develops as sand or silt particles are deposited by flowing water or wind. They measure in meters or yards.

Sedimentary rocks are layered, or stratified. If some layers are more resistant than others, they will weather into alternating ledges and slopes.

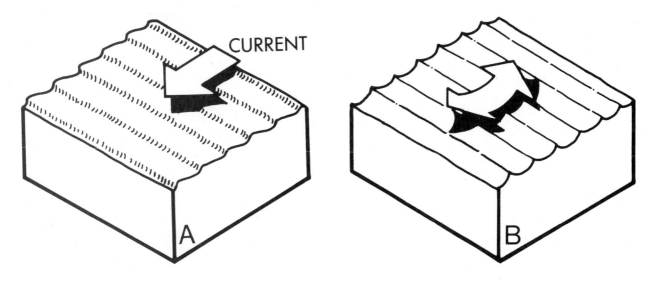

CURRENT

Ripple marks, measured in centimeters or inches, are sometimes found in sedimentary rocks.
A. Current-formed ripple marks have gentle upstream slopes and short, steep downstream faces.
B. Oscillation ripple marks, formed as water surges back and forth, are more symmetrical.

Water flowing from right to left built a miniature delta over older sand. Sand deposited on the steeply sloping downstream face will be cross-bedded.

• **Metamorphic rocks** form from preexisting rocks that are acted upon by intense pressure and/or heat. Their grains may simply fuse together, in which case their origins are not hard to recognize. Or they may be completely recrystallized, almost turned into new magma, so that it's quite difficult to deduce their original nature. In some, for instance, swirly dark and light banding is all that is left to suggest that they may once have been stratified. But whether they were sandstones or siltstones, or even lava and ash flows, cannot always be determined. Some metamorphic rock seems to grade into intrusive igneous rock, with the heat and pressure that caused recrystallization actually melting part but not all of the rock.

Intrusive igneous rocks come in masses of various shapes and sizes. The largest intrusions are **batholiths**, great bodies of granite or other equally coarse-grained rock. **Stocks** are smaller than batholiths, but like batholiths have no known bottoms. Many stocks are batholith off-shoots, poking up like fingers stretching toward but never reaching the surface.

Sheets of intrusive igneous rock may also be sandwiched between other rocks as **sills**, which parallel the layering of surrounding rocks, and **dikes**, which cut across surrounding rock layers. Both sills and dikes are crack-fillers, either injected with enough pressure to force older rocks apart, or flowing into cracks opened by shrinkage or other processes. They may branch from the feeders or conduits of volcanoes, and have the same chemical and mineral composition as the lava erupted at the surface. The intrusive rocks of stocks, sills, dikes, and **volcanic necks** usually harden more rapidly than do those of batholiths, so they are not as coarsely crystalline. A common pattern includes fairly large crystals of one or two minerals suspended in a much finer matrix—making the rock a **porphyry**.

All rocks are made of **minerals**, naturally occurring substances that are known to have definite chemical make-ups. Minerals commonly can be recognized by their color, hardness, and charac-

Sedimentary rocks may exhibit mud cracks formed as mud dries out and shrinks. These modern mudcracks formed in a drying desert puddle.

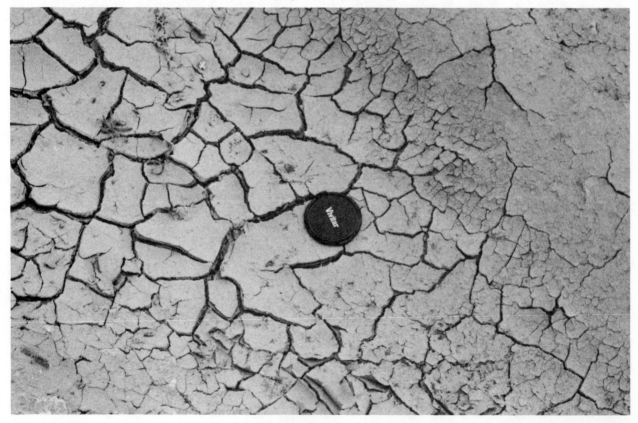

COMMON IGNEOUS ROCKS

INTRUSIVE IGNEOUS ROCKS					VOLCANIC EQUIVALENT		
	NAME	COLOR	COMP.	MINERAL MAKE-UP	NAME	FLUID.	OCCURRENCE
	GRANITE	light gray to white		mostly quartz and feldspars, with minor mica and/or hornblende	RHYOLITE		fine, pale ash and tuff
GRANITE	QUARTZ MONZONITE, GRANODIORITE	light gray	↑	mostly feldspars and quartz, with mica and hornblende	RHYODACITE	↓ increasingly fluid when molten	light gray ash or tuff, some volcanic domes
GRANITE	QUARTZ DIORITE	medium gray	increasingly silicic	mostly feldspars, with quartz, mica, and hornblende	DACITE		gray lava of volcanic domes, ash or tuff, pumice
	DIORITE	medium gray	↑	feldspars and quartz with abundant dark minerals	ANDESITE	↓	short, thick lava flows, volcanic ash or tuff
	DIABASE	dark gray to black		mostly dark minerals such as pyroxene, some plagioclase and quartz	BASALT		fluid black and dark gray lava flows, cinders

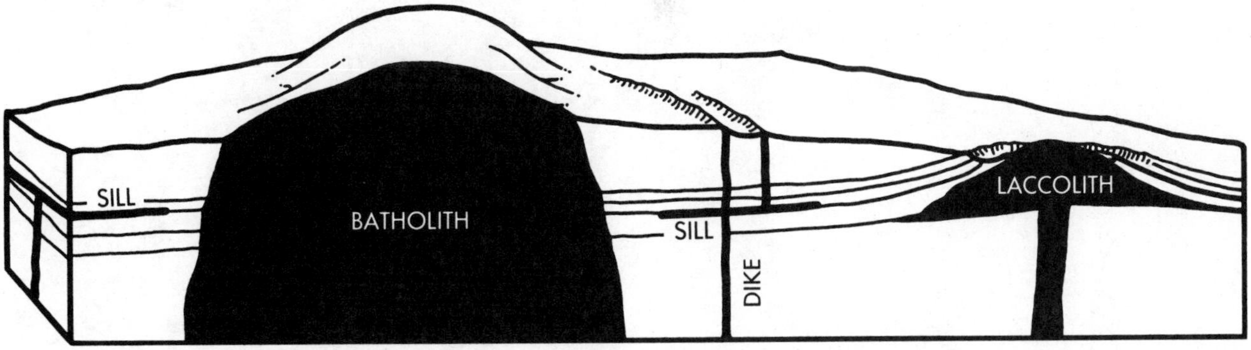
SILL
BATHOLITH
SILL
DIKE
LACCOLITH

Rising from the depths, magma that cools below the surface becomes intrusive rock. Intrusions come in many shapes and sizes. Later bared by erosion, their hard rocks stand out as mountains, hills, and ridges.

teristic ways of crystallizing or of breaking. You probably already have a speaking acquaintance with a number of minerals: quartz, mica, native gold, gemstone minerals like ruby and diamond, semiprecious minerals like turquoise, opal, and agate, and a very abundant mineral called water in its liquid phase, ice or snow in its solid or crystallized phase. Many beautiful minerals can be found in the Southwest, but since collecting of any kind is prohibited in national parks and monuments, minerals won't receive much attention in this book. By identifying one or two common minerals in each kind of rock, geologists refine rock descriptions, as, for instance, when they speak of **quartz dacite** or **biotite schist**. By far the most common of the rock-formers are quartz and members of the feldspar group. Calcite is the chief component of limestone.

VI. Geologic Dating

Geologists are fascinated with time—time in large doses. Geologic time goes back to the creation of the Earth about 4.6 billion years ago. But the rocks of the Earth can only be studied back to about 3.8 billion years, when the oldest rocks now known were formed. It is of course highly likely that they were preceded by even older rocks: The long history of the Earth involves constant recycling of Earth materials. The senior rocks in our southwestern ranges came into existence well over 2 billion years ago.

Two billion years: Time enough for major changes on the face of our planet. Time enough for the drifting of continents, their breaking apart, their coming together again only to break apart once more. Time enough, too, for the crea-

Molten magma squeezing into cracks and joints cools to form dikes. Harder than surrounding rocks, dikes often erode into dragonlike crests up to 20 meters (60 feet) high.

tion of mountains and their wearing away, and for more mountains, and more wearing away, over and over again. Time enough for the birth of life and for the long evolutionary path to modern plants and animals.

Obviously, normal hours and days, weeks and years, don't mean much in such immensities of time. Geologists have learned fairly recently how to measure geologic time by measuring the slow decay of radioactive minerals in rocks (**radiometric dating***) or by relating natural rock mag-

---

* Radioactive minerals decay at constant rates, starting at the time the rock containing them cools and solidifies. By measuring relative amounts of radioactive minerals and their decay products, we can judge the age of the rocks.

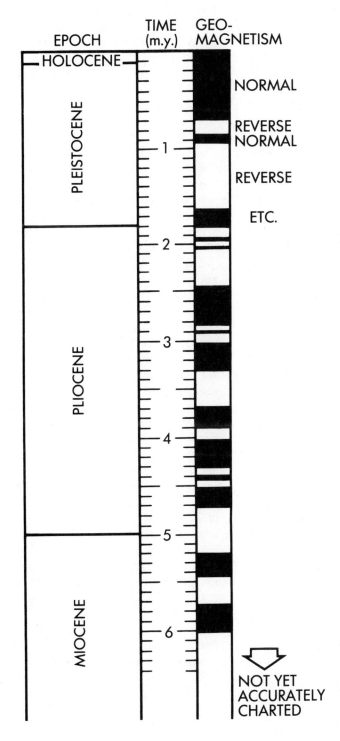

Many Cenozoic magnetic reversals have been accurately dated. Dark bands indicate normal magnetism as at present; light bands show periods when north and south magnetic charges switched places.

netism to the known history of reversals in the Earth's magnetic field, when the north and south magnetic poles switched their positive and negative charges (**paleomagnetic dating**). But not all rocks lend themselves to these techniques. It may be all we can do to say "This rock is older than that one," or "This one is much younger," as could be done with earlier methods of dating. Time was when dating methods relied on two things: the positions of rocks relative to each other, and fossils of known relative age preserved in rocks. These indicators worked reasonably well for sedimentary rocks, not so well for igneous rocks; the newer methods work best for igneous rocks.

Another type of dating, applicable only to quite young, quite recent deposits, relies on variations in the widths of tree rings. This method, used a great deal by southwestern archaeologists to date prehistoric ruins and artifacts, is equally useful for dating volcanic eruptions, floods, and other events that preserved trees or fragments of wood.

A firm tenet of geology (with exceptions, like most scientific "laws") says that in undisturbed sedimentary strata (rock layers), the oldest are at the bottom, the youngest on top. (This "law" also applies to layered volcanic rocks.) New, younger rock layers are laid down on top of old ones. Dating by relative position doesn't tell us exactly how long ago a given unit formed, but it does tell us that any one unit is younger than rocks below it, and older than rocks above—as long as the rocks are in approximately their original orientation.

As geologists began to study rock units older and rock units younger—and this was happening in a time when evolution of plant and animal life had not yet been demonstrated—they discovered that certain recognizable fossils could be found in certain recognizable layers, whether those layers were still horizontal or had been tipped up on end or even flipped over completely. When Charles Darwin opened the window of our present understanding of evolution, these earlier findings fitted into his theory as both evidence and verdict: Life has evolved; successive fossil-bearing layers illustrate this evolution; evidence of evolution can be used to define the ordered sequence of rock layers and their succession in the long march of geologic time.

Long before the advent of radiometric or paleomagnetic dating, geologists had devised a worldwide system of relative dating, a sort of geologic calendar, with special names for the months,

weeks, and days of geologic time. These names, made more exact by the new dating techniques, are still used by geologists today. In this calendar, the largest units, the geologic months, are called **eras**. Eras are divided into **periods**, and periods into **epochs**, as shown in the table. (Note that epoch names are given here for the Tertiary and Quaternary Periods only, as applicable in this volume.) The table also shows abbreviations for periods and epochs commonly used on maps and diagrams—a sort of geologic shorthand—and their approximate ages in millions of years.

Because the whole field of geologic dating is evolving rapidly, the dates on this table may differ by a few million years from those of other authors, as well as from those of earlier volumes in this series. The usefulness of the era, period, and epoch names, however, does not change.

Theoretically, all rocks—igneous, sedimentary, and metamorphic—can be fitted into the time scale, even though some rock units, lacking either fossils, radioactive minerals, or the iron minerals necessary for studies of magnetization, can only be dated relative to those around them.

Boundaries between the time divisions are not arbitrary. The era divisions were meant to indicate three stages in the development of life, from "early life" (Paleozoic) to "middle life" (Mesozoic) to "recent life" (Cenozoic). These three eras are also popularly called the Age of Fishes, the Age of Reptiles, and the Age of Mammals. When the Precambrian Era was first defined, life was thought not to have existed before Cambrian time. We now know that both plants and animals *did* exist then, in highly complicated forms. The ancestors of all our major groups (as well as some other groups with no modern descendants) had already evolved, but did not secrete the hard shells and skeletons likely to be preserved as fossils.

Eras and periods as first defined were thought to reflect very large breaks in the continuity of the rock record and the otherwise orderly record of life. Now we realize that the rock record is more or less continuous; gaps in the record in one area may be filled in somewhere else. Worldwide standards have been or are being adopted to define each era and period boundary; many epoch names, however, are applicable on only one continent.

There do seem to be sudden breaks in the orderly progression of plant and animal life. Some groups seem to go on forever, while others gradually or abruptly become extinct. Extinctions have puzzled geologists for decades. A possible

GEOLOGIC TIME

ERA	PERIOD	EPOCH	AGE IN YEARS
CENOZOIC Age of Mammals	QUATERNARY Q	HOLOCENE Q	
			— 10,000 —
		PLEISTOCENE Q	
			— 2 million —
	TERTIARY T	PLIOCENE Tp	
			— 5 million —
		MIOCENE Tm	
			— 24 million —
		OLIGOCENE To	
			— 37–38 million —
		EOCENE Te	
			— 55–57 million —
		PALEOCENE Tp	
			— 63–66 million —
MESOZOIC Age of Reptiles	CRETACEOUS	K	
			— 138–144 million —
	JURASSIC	J	
			— 205–208 million —
	TRIASSIC	℟	
			— 240–245 million —
PALEOZOIC Age of Fishes	PERMIAN	Pm	
			— 286–290 million —
	PENNSYLVANIAN	P or ℗	
			— 320–330 million —
	MISSISSIPPIAN	M	
			— 360–365 million —
	DEVONIAN	D	
			— 408–410 million —
	SILURIAN	S	
			— 435–438 million —
	ORDOVICIAN	O	
			— 500–505 million —
	CAMBRIAN	Ꞓ	
			— 570 million —
PRECAMBRIAN P-Ꞓ	ORIGIN OF LIFE		2.5 billion
	ORIGIN OF EARTH		4.6 billion

answer to the puzzle is now stirring up dust in the geologic world: Concentrations of the element **iridium**, which is rare on the Earth's surface but abundant in meteorites and in the Earth's mantle, occur right at the Cretaceous-Tertiary boundary. They suggest to many geologists that the extinction of the dinosaurs and many other life forms at the end of Cretaceous time may have been brought about by the impact and explosion of extraterrestrial bodies—perhaps comets or large iridium-rich asteroids—in meteor showers that may have lasted for thousands of years. As dust thrown up by the impacts and explosions spread out in the upper atmosphere, the world would have become shadowed and cold. Plants would have withered and died, their demise interrupting the food chains of many animals—the same pattern that has led scientists to warn the world about the possibility of a "nuclear winter." In addition, shock heating of the atmosphere by asteroid impact could have created nitrous oxide, which in turn could have combined with water to form nitric acid rain deadly to plants and animals.

Other geologists point out that iridium layers—and they have now been found at other earlier-recognized boundaries, corresponding with other extinctions—may have been brought about by unusually intense bouts of volcanism, which also might have heated up the atmosphere and thrown quantities of dust into the stratosphere. A third idea combines both concepts, stating that large asteroid impacts might trigger increased volcanism.

Boundaries between periods and epochs developed, over time, as geologists began to recognize lesser breaks in the rock and fossil records. Breaks in the rock record are called **unconformities**; they commonly take the form of old erosion surfaces that bevel or channel older tilted or folded rock layers, with younger rock layers deposited on top. Breaks in the fossil record may be due either to times of widespread uplift and erosion, to less significant asteroid or comet collisions, or to environmental changes. Extinctions great and small seem to occur in more than the usual abundance about every 26 million years, a periodicity that suggests an astronomical rather

In unconformities, younger rocks overlie the eroded surface of older rocks. Not yet consolidated into rock, this coarse gravel overlies much older limestone and the dike which cuts through it.

than an Earth-born cause, with perhaps passing stars deflecting comets or clusters of asteroids into the Earth's orbit.

VII. Maps and Figures

Ever since the days of heavy, oversize cameras and long exposures of big glass-plate negatives (along with mobile darkrooms carried on the backs of patient packhorses), photographs have been an integral part of geologic literature. However, maps, diagrams, and sketches sometimes give a better understanding of basic geologic patterns than do photographs. In this volume both are used, the photos largely to illustrate the text, the maps and diagrams to illustrate broader geologic patterns or covered-up geology not visible at the surface.

Geologic maps show specific, named, easily recognizable rock units called **formations** (or

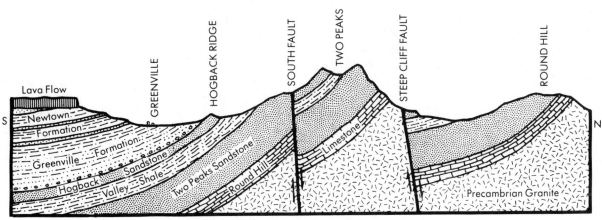

Sample Geologic Cross-Section

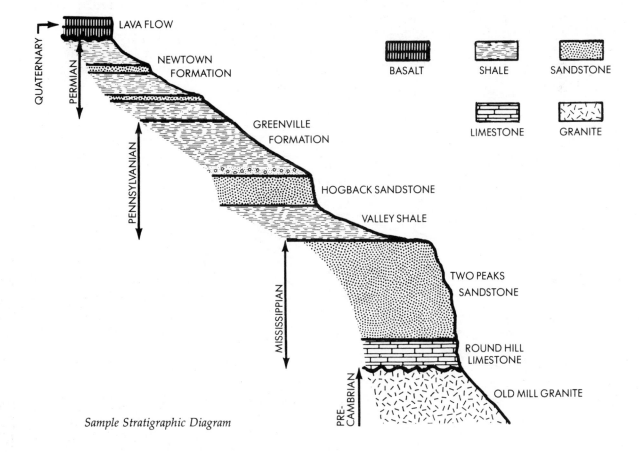

Sample Stratigraphic Diagram

groups of formations) that occur either at the surface or just under loose surface soil and rock debris. **Cross sections** diagrammatically slice open mountain ranges or other parts of the Earth's crust to give a picture of what geologists deduce is below the surface. **Block diagrams** combine cross sections with perspective views of surface features, showing what a particular block of the Earth's crust would look like if it could be lifted out and away from its normal surroundings.

Geologic maps are the prime product of most geologic field research. They represent weeks and months of slow, patient work by geologists who plot **outcrop** after outcrop, formation after formation, wherever rocks are exposed at the surface or within digging distance below the surface. A good geologic map, in color, carefully interpreted, gives a coherent picture of the geology of any one area. The U.S. Geological Survey (USGS) has geologically mapped much of the United States in considerable detail. Survey maps of 7 1/2- and 15-minute quadrangles, which cover areas of about 11 × 14 kilometers (7 × 9 miles) and 22 × 28 kilometers (14 × 18 miles), can be obtained from USGS map offices in Reston, Virginia; Denver, Colorado; and Menlo Park, California. Geologic maps of larger areas—whole states and in some cases individual national parks and monuments—are also available. Park and monument geologic maps, where they exist, can usually be purchased at visitor centers.

These maps show to the practiced eye not only the formations and groups but faults, anticlines, synclines, and the **dip** (downward slant) of rock layers. A much-used geologic term, dip is measured in degrees below the horizontal, in the direction of the greatest slope of a rock. On maps it is represented by the dip symbol ⌐28 in which the shorter line points in the direction of dip and the number, where used, shows the degrees. Interpretation of geologic maps is a skill well worth learning. The few simplified maps in this book are but a beginning; very little detail can be shown on such a small scale.

Cross sections and block diagrams are usually easier to understand than maps, and are good ways to show geologic features at a glance. They are made from maps, or, where rock layers are well exposed, from the rocks themselves. Block diagrams, such as those used earlier in this chapter to illustrate folds and faults, are particularly useful for interpreting the geologic and topographic changes that come with faulting, folding, uplift, and subsequent erosion. In both cross sections and block diagrams, the vertical dimension may be exaggerated in order to show the succession of rock layers more clearly. Such vertical exaggeration also accentuates folds, steepens faults, and intensifies the ruggedness of surface profiles.

Another type of illustration, particularly suited to areas with many layers of sedimentary rock, is the **stratigraphic diagram**—a picture of rock layers exposed in many outcrops, perhaps over a wide area, as they would appear if piled on top of one another in the order in which they were originally deposited: oldest at the bottom,

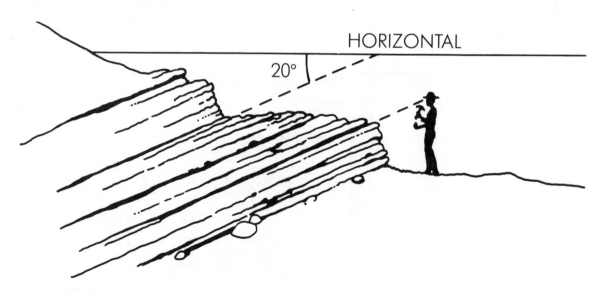

*The **dip** of tilted rock layers is measured in degrees below horizontal. On maps, the symbol ⌐ shows direction of maximum dip (short line).*

Ancient metamorphic rock shows patterns of distorted layers that may once have been sea-floor sands and muds.

youngest on top. Some stratigraphic diagrams also show how these rock layers weather and erode—some as slopes, some as ledges and cliffs. Under natural conditions, though, individual rock layers may not everywhere erode in exactly the same way, so one must make allowances when matching up scenery with stratigraphic diagrams.

VIII. Geologic History

With these basic concepts in mind, we are able now to look at the geologic history of the southwest deserts and the way they fit into the geologic scheme of North America as a whole. Let's take their history era by era.

Precambrian Era. The oldest rocks in the Basin and Range deserts are Precambrian: gneiss and schist formed more than 2 billion years ago. The basic material of these rocks was formed even earlier, perhaps as alternating sequences of sandstone and shale, or of volcanic lava and ash. In the Earth's recycling system the sedimentary and volcanic rocks, changed by heat and pressure, became metamorphic rocks. They were intruded some 1.8 billion years ago by granite. In just a few areas, as in the Panamint Range west of Death Valley and in the region around Tonto National Monument in Arizona, the gneiss, schist, and granite are overlain by younger Precambrian rocks, much less altered and clearly recognizable as sedimentary in origin. There are

layers of sandstone (much of it altered to quartzite), shale, and some conglomerate, the kinds of sedimentary rocks likely to be deposited in shallow sea water fairly close to land. Bridging the gap between Precambrian and Paleozoic Eras, these rocks suggest that late in Precambrian time the western margin of the continent—a low, flat margin receiving sediments from the continental interior—lay somewhere between the Sierra Nevada and the present Rocky Mountain region.

Precambrian uplift and mountain-building were followed in most parts of North America (as well as in Europe and Asia) by millions of years of erosion, a general beveling of the land, and then by the advance of a Paleozoic sea. Except in areas with younger Precambrian sedimentary rocks, the contact between Precambrian and Paleozoic rocks is sharp, an unconformity between the very old metamorphic and igneous rock, and much younger Cambrian sandstone. Unconformities, incidentally, are quite common, with time gaps great and small, throughout the rock record. More than half of our pages of stone are gone forever.

Paleozoic Era. During the whole of the Paleozoic Era the beveled continent lay quiet. For 300 million years, it warped a little, gently sank and rose, allowing the western sea to come and go across it. But that is all. In much of western North America there was neither volcanism nor intrusion of magma nor mountain-building of any kind during this entire time.

Cambrian sedimentary rocks of this region are as much as 5000 meters (16,000 feet) thick, as shown by the tilted rocks of several ranges. Studies show that the lowest Cambrian layers, those that lie directly on Precambrian rocks, become younger eastward, as if the Cambrian sea in which they were deposited gradually crept eastward across the flattened, slowly subsiding continent. The eastward thinning of Cambrian sediments extends well beyond the limits of the Basin and Range area: They are substantially thinner—about 300 meters (1000 feet)—in the Grand Canyon, and less than 100 meters (300 feet) in the Great Lakes region.

Cambrian sandstone, siltstone, and limestone contain an abundance of fossils, for shells had evolved, and shell-secreting marine life—trilobites, brachiopods, and sponges—flourished in the warm, shallow sea.

Ordovician, Silurian, and Devonian rocks are rare in the southwest ranges. Where they do occur, they are predominantly limestone, indicating once more a fairly warm, quiet marine environment. They, too, contain fossils, including many types of brachiopods, corals, and trilobites, as well as numerous mollusks and the bony armor plates of early fishes—the first of the vertebrate line.

The Mississippian Period is well represented over almost all of the Southwest by massive gray limestone, a unit that, under one name or another, extends over a good deal of western United States. Rich in fossil horn corals, brachiopods, crinoids (sea lilies), and other forms, it tells us of a long stable period when the western sea encroached far across the continent. Late in Mississippian time, however, parts of this area rose above the sea, and erosion removed some of the strata. The climate was tropical, and an irregular surface with dissolved-out caverns and collapsed **sinkholes** formed on some of the Mississippian limestone.

In Pennsylvanian time, uplift of the Ancestral Rocky Mountains to the north and northeast and their subsequent erosion brought contributions of clay, silt, and sand to parts of this region.

In Permian time the sea once more swept in from the west and to some extent from the south. In it, more fossil-bearing marine limestones were deposited. The climate during all of Paleozoic time seems to have been warm and tropical; North America had not yet drifted to its present Northern Hemisphere position. In parts of the present Southwest, shallow bays were sites of great limestone reefs similar in many respects to tropical coral reefs today.

Mesozoic Era. The picture changed during the Mesozoic Era, the time of the dinosaurs, as forces opening up the Atlantic Basin began to flex and crumple parts of the continent. The great western sea drained away. In places, huge balloon-shaped masses of molten rock pushed upward and slowly cooled and solidified. West of our Basin and Range Province, granites of the Sierra Nevada also worked their way upward; the ancestral Sierra Nevada held back Pacific moisture in a foretaste of the climate control exercised by the present range. Toward the end of Mesozoic time, and extending into the Cenozoic Era, the present Rocky Mountains were born—far northeast of our area but providing headwaters for the two largest desert rivers, the Rio Grande and the Colorado River. Nature's response to uplift is erosion, and erosion of the rising Rockies brought sand, gravel, and clay into the southwest lowlands, spreading them far across the earlier surface. Finally, as the Atlantic Basin widened and the continent drifted westward, the Mesozoic Era came to a close, its end perhaps punctuated by great meteor showers, the impact of many asteroids, and the long night of airborne dust.

Cenozoic Era. The oldest Cenozoic deposits in the Basin and Range area were formed about 30 million years ago at the bottom of the just-born Rio Grande Rift. These deposits, river-carried sand and mud as well as volcanic ash, contain the fossilized remains of both land plants and mammals. The mammals—horses, hyenalike dogs, mastodons, peccaries, and ancestors of today's camels—lived in a moist savannah-type environment, a grassland considerably lusher than today's desert. Soon the region would become more arid as the faraway Sierra Nevada, edged by faults, rose once again to challenge winds from the west.

At about the same time, much of the territory west of the Rio Grande Rift was also changing. Geologists have come up with more than one explanation for these changes. Some think the crust was shortened and North America narrowed as large-scale thrust faulting took place. Others believe that the initial faulting stretched or widened the crust, perhaps by as much as 400 to 500 kilometers (250 to 300 miles). Faulting of both these types slides an upper plate across a lower plate, the movement taking place on nearly horizontal faults, as if you laid one hand on top of the other and then slid them farther over each other or pulled them apart. Geologists call these movements **compressional** (pushed-together) and **extensional** (pulled-apart) faulting. In parts of the

Southwest, both types of low-angle faulting took place—the pushed-together type 70 to 60 million years ago, the pulled-apart type around 30 million years ago.

Another round of faulting, starting 15 to 11 million years ago (and in some places still going on), raises less disagreement among geologists. Called the Basin and Range Disturbance, this second faulting was definitely controlled by ten-

sion, a pulling apart. But this time tension produced steep, in many cases nearly vertical, faults that today edge the basins and ranges. Ranges rose as blocks between pairs of faults; basins sank as tension pulled the fault blocks apart. The two long slivers that are the Rio Grande Rift and the Salton Graben once more subsided between their bordering faults, as they continue to do, sporadically, today.

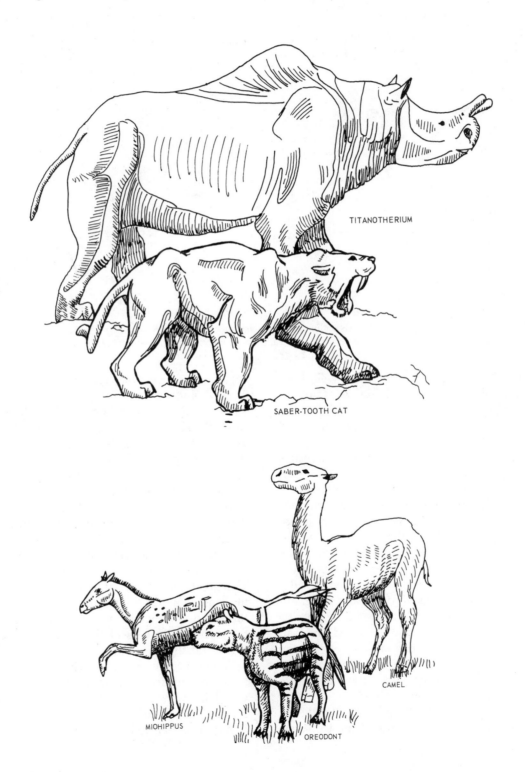

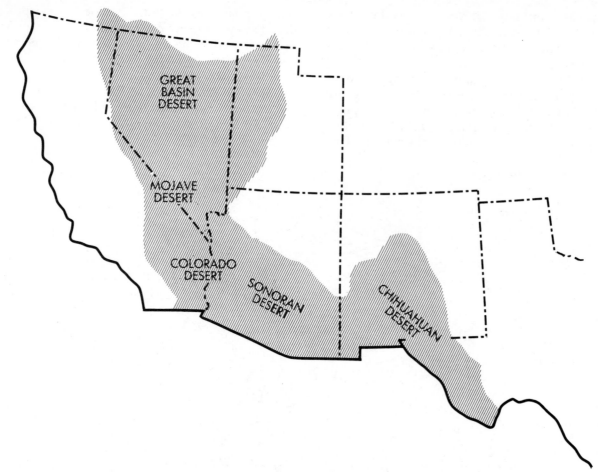

Five major deserts, differing in elevation, climate, and biological characteristics, make up the Basin and Range Province. The Colorado Desert, near the Colorado River, is commonly considered part of the Sonoran Desert.

Volcanoes raised their fiery heads as well, both before and after Basin and Range faulting. Molten magma squeezed upward along fissures and faults to create lava flows, cinder cones, and some sizable cone-shaped stratovolcanoes. Great volcanic explosions covered the land with thick blankets of volcanic ash, and led to collapse of some of these volcanoes and development of huge basin-shaped **calderas**. (An old rule-of-thumb about Basin and Range volcanic rocks declares that if they are tilted they are Tertiary, and if they're flat-lying they're Quaternary. The trouble with this rule is that there are *lots* of exceptions!) Desert landforms continued to develop: great alluvial fans, broad bajadas, gaunt ranges, sandy desert washes. Many basins lacked external drainage and collected the salt deposits of playas and saline lakes—not just the table salt variety but also borax, saltpeter, gypsum, sodium carbonate, and other **evaporite** (concentrated by evaporation) minerals.

The climate changed abruptly in Pleistocene time, in harmony with the onset of the Ice Ages.

No glaciers ever reached the low deserts, but large alpine glaciers appeared in the Sierra Nevada and Rocky Mountains, and small ones clung to steep, shadowed valleys in a few of the highest desert ranges. Rain-heavy clouds surmounted the Sierra barrier, bringing to the desert rainy cycles in step with the pulses of great ice sheets farther north. Increased rainfall created lakes in undrained basins and established through drainage where there had been none before. Erosion intensified in the mountains, with debris deposited in the valleys. Grasslands and forests developed where today's cacti and Joshua trees grow. The great rift valleys, still deepening by sporadic movement along their bordering faults, received thousands of meters of sediment brought in by the two great rivers of the region, the Colorado River and the Rio Grande.

Then, little more than 10,000 years ago, long after man had crossed the Bering Strait and wandered his way down through Central and South America, the ice retreated, and the Basin and Range deserts dried out once more.

IX. Other Reading

In this book, as in others of this series, the parks and monuments are discussed in alphabetical order. In addition to the true Basin and Range areas, this volume includes three national monuments—Tonto, Montezuma Castle, and Tuzigoot—in a transition zone between the lowland deserts and the canyon country to the north. It also extends to Capulin Mountain National Monument at the edge of the Great Plains in northern New Mexico.

To find your way around in park and monument areas, refer to small maps distributed at entry gates by the National Park Service, or purchase more detailed topographic maps at visitor centers. Some visitor centers have free handout sheets describing the local geology. Many nature trail leaflets note geologic features. And some issues of the magazines *Arizona Highways* and *National Parks* contain passing references to the geology of specific parks and monuments.

The books and articles listed below are of a more general nature: Some address the geology of whole states or regions, or geologic processes such as volcanism or evolution of landforms. Many of these books are somewhat technical, but others are written for readers with little or no background in geology. All will in some way enrich your understanding of the geology of our parks and monuments. In addition to books listed here, there are many good geology textbooks, at both high-school and college levels; those published after 1975 are most likely to include the Theory of Plate Tectonics.

Angier, Natalie, 1985. "Did Comets kill the Dinosaurs?" *Time*, vol. 125, no. 18 (May 6): pp. 72–83.

Arizona Bureau of Mines and Mineral Resources. *Geologic Guidebooks to Highways of Arizona*. University of Arizona, Tucson.

Axelrod, D. I., 1981. *Role of Volcanism in Climate and Evolution*. Geological Society of America Special Paper 185.

Bullard, F. M., 1962. *Volcanoes: in History, in Theory, in Eruption*. University of Texas Press, Austin.

Chapin, C. E., and Elston, W. E. (editors), 1979. *Ash-Flow Tuffs*. Geological Society of America Special Paper 180.

Chronic, Halka, 1983. *Roadside Geology of Arizona*. Mountain Press, Missoula, Montana.

Coffman, J. L., von Hake, C. A., and Stover, C. W. (editors), 1982. *Earthquake History of the United States*. National Oceanic and Atmospheric Administration Publ. 41–1.

Colbert, E. H., (editor), 1976. *Our Continent: a Natural History of North America*. National Geographic Society, Washington, D.C.

Cowen, R., 1975. *History of Life*. McGraw-Hill Book Co., New York.

Decker, R. and B., 1981. *Volcanoes*. W. H. Freeman and Co., San Francisco.

Hamblin, W. K., 1985. *The Earth's Dynamic Systems*. Burgess, Minneapolis.

Hamilton, Warren, 1978. *Plate Tectonics and Man*. Reprinted from U.S. Geological Survey Annual Report, Fiscal Year 1976. U.S. Government Printing Office, Washington, D.C.

Jaeger, E. C., 1957. *The North American Deserts*. Stanford University Press, Stanford, California.

Kurten, B., 1972. *The Age of Mammals*. Columbia Press, New York.

Macdonald, G. A., 1972. *Volcanoes*. Prentice-Hall, Englewood Cliffs, New Jersey.

Marvin, U. B., 1973. *Continental Drift*. Smithsonian, Washington, D.C.

Nations, D., and Stump, E., 1981. *Geology of Arizona*. Kendall/Hunt, Dubuque, Iowa.

New Mexico Bureau of Mines and Mineral Resources. *Scenic Trips to the Geologic Past* (guidebooks to New Mexico highways). Socorro, New Mexico.

Oakshott, G. B., 1975. *Volcanoes and Earthquakes: Geologic Violence*. McGraw-Hill Book Co., New York.

Péwé, T. L. (editor), 1981. *Desert Dust: Origin, Characteristics, and Effect on Man*. Geological Society of America Special Paper 186.

Smiley, T. L., Nations, J. D., and others, 1984. *Landscapes of Arizona—The Geological Story*. University Press of America, New York.

Sullivan, W., 1974. *Continents in Motion*. McGraw-Hill Book Co., New York.

Vine, F. J., 1970. *Sea-Floor Spreading and Continental Drift*. Journal of Geological Education, vol. 18, no. 2.

Wilson, J. T. and others, 1972. *Continents Adrift. Readings from Scientific American*. W. H. Freeman and Co., San Francisco.

Wyllie, P. J., 1971. *The Dynamic Earth*. Wiley, New York.

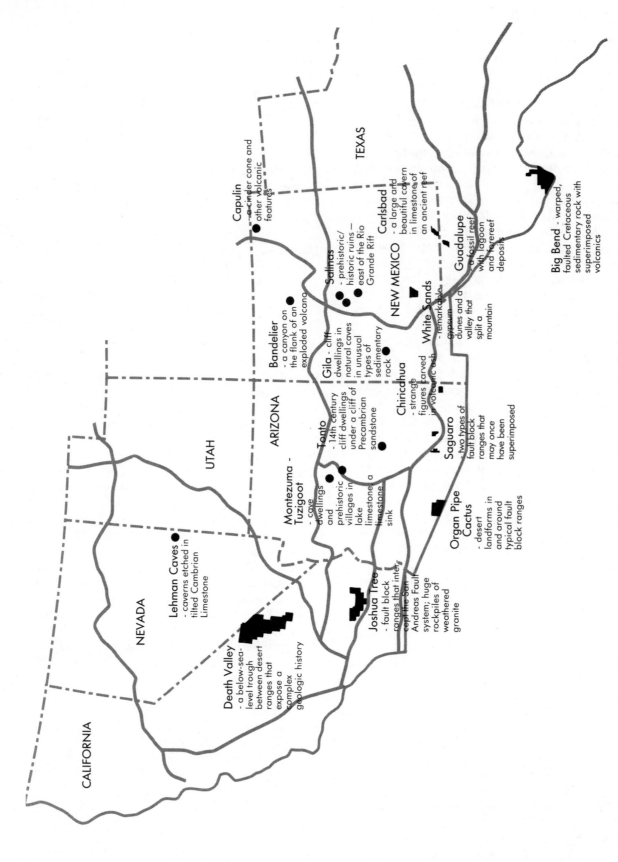

CALIFORNIA

NEVADA

UTAH

ARIZONA

Lehman Caves ●
- caverns etched in
tilted Cambrian
Limestone

Death Valley
- a below-sea-
level trough
between desert
ranges that
expose a
complex
geologic history

Joshua Tree
- fault block
ranges that inter-
cept the San
Andreas Fault
system; huge
rockpiles of
weathered
granite

Montezuma -
Tuzigoot
- cave
dwellings
and
prehistoric
villages in
lake
limestone a
limestone
sink

Organ Pipe
Cactus
- desert
landforms in
and around
typical fault
block ranges

Tonto
- 14th century
cliff dwellings
under a cliff of
Precambrian
sandstone

Bandelier
- a canyon on
the flank of an
exploded volcano

Gila - cliff
dwellings in
natural caves
in unusual
types of
sedimentary
rock

Chiricahua
- strange
figures carved
in volcanic ash

Saguaro
- two types of
fault block
ranges that
may once
have been
superimposed

Capulin
- a cinder cone and
other volcanic
features

Salinas
- prehistoric/
historic ruins
east of the Rio
Grande Rift

NEW MEXICO

Carlsbad
- a large and
beautiful cavern
in limestone of
an ancient reef

White Sands
- remarkable
gypsum
dunes and a
valley that
split a
mountain

Guadalupe
- a fossil reef
with lagoon
and forereef
deposits

TEXAS

Big Bend - warped,
faulted Cretaceous
sedimentary rock with
superimposed
volcanics

National parks and monuments of the Basin and Range region demonstrate many types of geologic features, all easily seen because of the dearth of vegetation.

PART 2.

THE PARKS AND MONUMENTS

Bandelier National Monument

Established: 1916
Size: 150 square kilometers (58 square miles)
Elevation: 1849 meters (6066 feet) at visitor center
Address: Los Alamos, New Mexico 87544

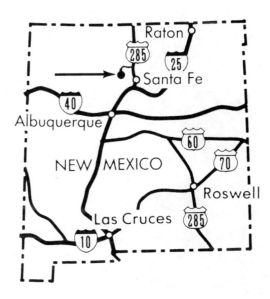

STAR FEATURES

• Ruins of prehistoric dwellings, in a peaceful vale surrounded by cliffs of tuff (volcanic ash) born in the violence of a great explosion.

• The nearby Rio Grande Rift and Jemez Caldera, whose histories explain the presence of this rock.

• An informative visitor program that includes slide shows, museum exhibits, and self-guide leaflets, as well as conducted tours and evening campfire lectures in summer.

SETTING THE STAGE

The ruins in Bandelier are set in Frijoles Canyon, one of the many cliff-walled gorges that cut through the Pajarito (pa-ha-REE-toe) Plateau, on the east slope of the Jemez (HAY-mess) Mountains, at the southern end of the Rocky Mountain chain. Here, where water, game, and rich soil came together, a prehistoric people built their homes. Behind the hard crust of the canyon ramparts the newcomers found the pink-gray volcanic rock soft enough to cut, scrape, and gouge with stone and bone tools. In it they shaped storage and sleeping rooms, probably by enlarging and connecting existing alcoves and cavities. Against the canyon wall they constructed rows of rooms, working sockets into the cliff to hold roof beams. And on the verdant valley floor they built a walled town, which they occupied from about 1300 to 1550 A.D., just before the arrival of Spanish conquistadors and settlers.

On close examination the canyon walls are not nearly as smooth as they look from a distance. The rock contains hard shards of volcanic glass and irregular lumps of other volcanic material, all of which give it a rough texture. It is composed of bubbly fragments of volcanic ash, of the composition of rhyolite, fused together into welded tuff. In places this tuff is quite weather-resistant, on the surface at least, but elsewhere erosion has worn through the hard exterior and hollowed out natural alcoves and overhangs. The harder outer shell is a form of case-hardening. Moisture from rain or snow sinks into this relatively porous rock and dissolves minute quantities of silica from the rock itself. Later, the silica-bearing moisture is drawn to the surface by warm sunshine and dry air. As the water evaporates it

Using tools of bone, horn, and rock, early inhabitants enlarged and decorated rooms within the Bandelier Tuff. Small round holes were sockets for roof beams of dwellings built in front of the cliff.

leaves its dissolved minerals behind to strengthen and harden the surface layer.

Numerous vertical cracks or joints, emphasized by weathering and erosion, cut through these rocks. Partly because of these vertical fissures, partly because of the hardened surfaces, and partly because volcanic fumes rising through the porous ash may have hardened parts of it into chimneys, the rock in some places erodes

into "tent rocks," conical forms more like tepees than tents.

Frijoles Canyon is in many ways typical of the desert canyons that drain southwestern mountain ranges. Its stream, fed by rain, snow, and springs in the Jemez Mountains, babbles across sand-floored vales cooled by shade trees. A short distance downstream its character changes. The canyon narrows and the stream plunges over two waterfalls. Canyon walls there are composed of dark lava flows from the Cerros del Rio volcanoes on the other side of the Rio Grande. Along the walls of this lower canyon can be seen many features of those flows—thin basalt layers, red-baked soil zones beneath them, vesicles (gas-bubble holes), and little green olivine crystals in the basalt. Unlike most streams emerging from desert mountains, Frijoles Creek rarely dries up. It joins the Rio Grande about 4 kilometers (2.5 miles) below the visitor center.

The geology at Bandelier is tied closely with the remarkable features of the Jemez Mountains and their central depression, the Jemez Caldera. The range is all that is left of a lofty volcano— once similar in shape and size to Fujiyama or Mount Hood—that gazed down in snow-capped splendor upon the Rio Grande Valley a million years ago. Those who wish to understand Bandelier should by all means also drive west about 25 kilometers (15 miles) on New Mexico

Highway 4 to see the circular, grass-floored caldera created by the volcano's collapse. Spanish settlers named it Valle Grande (Big Valley). In the opinion of many, the range and caldera themselves deserve national park or monument status and protection.

Just east of the Jemez Mountains, the Rio Grande flows southward through the long, slender graben of the Rio Grande Rift. Dropped between two bands of faults that stretch from central Colorado to southern New Mexico, this is one of the great rift valleys of the world. Its bordering faults reach all the way down to the upper mantle. To learn more about this rift and its relationship to Bandelier National Monument and the Jemez Caldera, let's take a look at the geologic history of this area.

GEOLOGIC HISTORY

Cenozoic Era. The faulting that created what is now the Rio Grande Rift began fairly late in

In Frijoles Canyon, thick layers of ash from the Jemez Volcano provided living space and building materials for a prehistoric people.

National Park Service photo.

Rain and wind have carved goblins and ghouls in the soft rock of the Bandelier Tuff.

Frijoles Creek carved its canyon in the welded tuff of a great volcanic explosion. Part of the rim of the Jemez Caldera appears in the background.

Cenozoic time, probably as recently as 30 million years ago. It was superimposed on older geologic structures related to the building of the Rocky Mountains and their ancestors, ranges that occupied about the same position as the present Rockies. Formation of the rift valley seems to have been controlled to some extent by the position of these older ranges. But the Rocky Mountains and their ancestors were formed by horizontal compression—a pushing-together that corrugated the crust—while the Rio Grande Rift was created by tension—a pulling apart. Some very basic changes must have taken place some-

where below the Earth's crust sometime between the building of the Rockies about 65 million years ago and the development of the Rio Grande Rift starting 30 million years ago. The rift may represent a line along which two giant "boils" in the mantle slowly roll apart, as they do along mid-ocean ridges. If this is so, the Rio Grande Rift may represent the very beginning of a new ocean basin, an earlier stage than that shown by the Red Sea Rift or the rift that separates Baja California from the Mexican mainland. (For a discussion of the latter, see Joshua Tree National Monument.)

Over many millions of years, faults that edge the Rio Grande Rift offered routes by which lava could reach the surface. Both Frijoles Creek (foreground) and the Rio Grande (background) have cut down through many layers of ash and lava.

Rift faulting was accompanied by both intrusion and volcanism, for the great pull-apart faults provided easy avenues for magma's upward flow. Both volcanic and intrusive rocks can be dated radiometrically, giving us a way to date development of the rift. Active movement along the faults was sporadic. Periods of activity, when fault movements and the earthquakes they engendered were frequent, alternated with periods of stability, when erosion attacked the highlands on either side of the rift, whittling down the mountains and carrying rock debris into the rift valley itself. This rift valley is really much deeper than it now appears, as it is largely filled in with rock debris. Total offset along the faults that edge it measures about 8000 meters (25,000 feet). And rifting is still going on: The entire valley is subject to minor earthquakes.

About two million years ago, in the area of Cerros del Rio just southeast of Bandelier National Monument, magma surged to the surface to erupt as lava flows. These flows were wide-

National Park Service photo.

Near and below the falls, Frijoles Canyon is walled with dark volcanic rocks derived from the Cerros del Rio volcanic field.

spread, and established the broad base on which the Jemez volcano would later stand, a many-layered base whose rocks can be seen today along the Rio Grande or in the walls of Frijoles Canyon along the Falls Trail. Small green crystals of olivine in the basalt tell us that the lava came from the mantle, deep below the Earth's crust.

Like the Cerros del Rio basalt flows, the Jemez volcano was a product of the Rio Grande Rift: The magma of which it was built rose through the great faults that edge the rift. But unlike the earlier lavas, this magma was so thick and sticky that it either formed short, thick lava flows or burst out with explosive eruptions that sent towering clouds of volcanic ash into the air. In repeated eruptions, lava flows and volcanic ash built up a many-layered volcano, a stratovolcano. By about 1.1 million years ago the Jemez volcano reached its maximum size.

Then close to a million years ago, in two sudden climax eruptions, the volcano spewed out incredible quantities of volcanic gas, ash, pumice,

Finely layered rock in canyon walls near Upper Falls resulted from steam explosions occurring as water came in contact with hot lava. Here, several volcanic bombs, thrown out by the explosion, fell into the fine mud before it solidified.

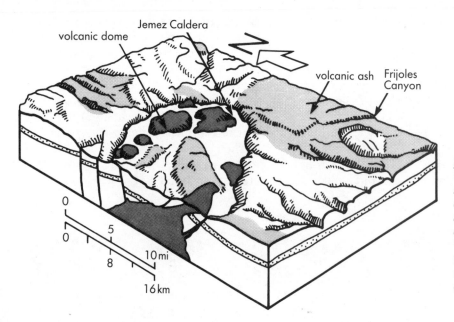

volcanic dome

Jemez Caldera

volcanic ash   Frijoles Canyon

0

0   5

8   10 mi

16 km

Jemez Caldera, the Valle Grande of Spanish settlers, resulted from a catastrophic volcanic explosion and subsequent collapse of the Jemez Volcano. Domes within the caldera formed after the main explosion. (After USGS.)

and broken rock—more than 200 cubic kilometers (50 cubic miles) of rock material, about 100 times that released by the 1980 Mount St. Helens eruption! Some of the volcanic ash rose skyward in dark, roiling, gargantuan clouds, from which it rained down to cover the landscape with choking dust. Some of it drifted on high-altitude winds, coming to earth as far afield as Iowa, Oklahoma, and Texas. Some was propelled downslope from the summit crater in dense, fast-moving, super-heated ash flows buoyed up by their own heat and by the expanding gases within them. As they came to rest on the volcano's slopes, these ash flows fused into welded tuff. That of the second climax eruption became the light pinkish gray Bandelier Tuff that now forms the walls of Frijoles Canyon.

Much of the Bandelier Tuff represents a single cooling unit—one thick layer of volcanic ash that may have taken many decades to cool completely. As such, the unit is welded more tightly in its center than near its upper and lower margins, where it was exposed to air or to cooler underlying rock. Differences in degrees of welding have regulated the mesa-and-canyon topography here. Most of the soft upper part of the tuff unit has eroded away completely, so that the hard central part forms the mesa surface. As streams carved channels through the hard part and into the soft lowest part of the unit, they undermined the harder unit, which fell away, leaving the steep cliffs that now wall Frijoles Canyon.

The two climax explosions removed vast quantities of gassy magma from the magma chamber deep down below the volcano, and in so doing left a yawning subterranean void directly underneath the main mountain mass. No longer supported from below, the central part of the mountain, ringed by fractures, collapsed—perhaps little by little, perhaps all at once—into the magma chamber void. Its collapse produced, at the surface, an almost circular, cliff-ringed pit 22 kilometers (14 miles) across, the Valle Grande or Jemez Caldera.

More than twice the diameter of fabled Crater Lake in Oregon, and early in its history also lake-filled, this caldera saw some further changes. The floor was domed up a little by continued pressure from below, and small, nonviolent eruptions of thick, sticky lava rose through the ring of fractures at the edge of the caldera, building a circlet of steep-sided lava domes—now tree-covered—within the Valle Grande.

After the big collapse, erosion resumed its eternal task, carving rills, then ravines, and then canyons into the welded tuff that flanks the defunct volcano. Eventually an outlet was excavated through the southwest caldera wall, and the lake drained. Nearly a million years after the great explosion, in a sheltered southeastern canyon, an agricultural people found security and sustenance. Using materials at hand—the welded tuff of the great explosion—they built the homes and villages whose ruins now mark the valley floor.

OTHER READING

Eichelberg, J. and Heiken, G., undated. *Geology of Frijoles Canyon: a Field Trip on the Falls Trail.* Los Alamos Scientific Laboratory.

Big Bend National Park

Established: 1944
Size: 2865 square kilometers (1107 square miles)
Elevation: 533 to 2388 meters (1750 to 7808 feet)
Address: Big Bend National Park, Texas 79834

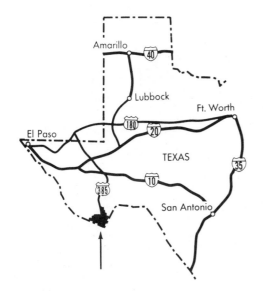

STAR FEATURES

• A section of the Basin and Range Province in which corrugated and faulted Cretaceous sedimentary rocks are hosts to volcanic outpourings and igneous intrusions.

• Landforms that relate to a varied sedimentary history as well as to faulting, intrusion, and volcanism. Today's desert processes are superimposed on features developed during times of wetter climate.

• The Chisos Mountains and Santa Elena, Mariscal, and Boquillas Canyons, with varied geology and spectacular scenery in each.

• An excellent interpretive program that includes museum displays, guided nature walks, workshops, movies, evening slide programs, and roadside exhibits along a highway expressly planned for viewing geologic features.

See color pages for additional photographs.

SETTING THE STAGE

Most of Big Bend National Park is low-lying desert. More or less in the center of the park rise the rugged Chisos Mountains. Many smaller mountains, mesas, hills, and ridges project, some of them quite abruptly, above the desert lowland. In the extreme west and along the northeast border, tall cliffs bound the highlands of Mesa de Anguila and Sierra del Carmen. The Rio Grande has cut deep canyons through these two highland areas, as well as through Mariscal Mountain at the apex of the river's "Big Bend."

To the first geologists who came here the area's geology seemed extremely complex. But careful study has shown a basic pattern. A broad expanse of gently warped and locally faulted Cretaceous sedimentary rock underlies almost all of the park. Prominent fault zones edge the eastern and western highlands, and embrace a large central sunken block dropped down along the fault zones. Superimposed on this pattern are many igneous intrusions that cut through the layered rocks, as well as broad lava flows and drifts of volcanic ash. Faulting, folding, and vol-

The Chisos Mountains, near the center of Big Bend National Park, consist of thickly piled volcanic flows and several small intrusions. Like many desert ranges they rise quite abruptly from the desert floor.

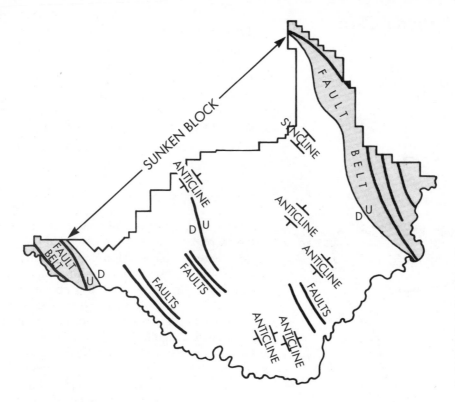

Big Bend's sunken central block is bordered on the northeast and southwest by faulted, uplifted ranges. Several anticlines and faulted synclines ripple the surface of the sunken block. These structures appeared before alteration of the landscape by intrusions and volcanic activity.

canism are responsible for smaller hills, buttes, and mesas that jut above the general surface of the sunken block.

Many typical desert features show up in this park. Alluvial fans reach out in all directions from the Chisos Mountains, partly covering the eroded mountain pediment, a beveled surface carved in the rock of the mountain itself. Pebbles in the gravels that surround the mountains range in age from Paleozoic through Cretaceous and into Tertiary, and thus represent a time span of more than 500 million years.

Farther from the mountains, the broad alluvial fans merge into typical desert bajadas, which in turn merge with the adjacent valley fill. Elsewhere, barren cliffs jut skyward; from them, rocks loosened by rain, wind, tree roots, small animals, or an occasional hard frost tumble to rocky talus slopes below. Broad, normally dry washes bear the braided imprint of intermittent streams on their sandy beds. Here also are many examples of desert pavement, an armorlike layer of pebbles and cobbles winnowed by the wind. Desert varnish, a thin coating of manganese and iron oxides, darkens rock surfaces. Badlands develop as areas underlain by soft volcanic ash are eroded.

Features of stream erosion and deposition are visible all along the Rio Grande. The river cuts steep-walled canyons through relatively hard

Mesozoic sedimentary rocks, yet broadens its valley and wanders easily where the bordering rocks are less resistant. In broad parts of the valley, sand bars and streamside dunes mark the river's course. Floods leave their marks along the Rio Grande's banks and along such large tributary streams as Tornillo and Terlingua Creeks, marks that last for many succeeding non-flood years.

GEOLOGIC HISTORY

Paleozoic Era. Big Bend's history—that portion of it that we can decipher—began in Paleozoic time when some of the area lay under a long trough-like arm of the sea. Layers of marine limestone and sandstone deposited at that time, all of them originally flat-lying, contain shells of numerous sea animals—brachiopods, corals, and other invertebrates—that enable geologists to pin down the age of the rocks.

At the end of the Paleozoic Era, movements of the crust raised and buckled the marine deposits to form an ancient mountain range. Because of this uplift and folding, erosion during early parts of the Mesozoic Era erased most signs of these mountains and of their Paleozoic rock layers. A few telltale fragments, however, remain in thin fault slices near Persimmon Gap at the north entrance to the park.

Mesozoic Era. During much of Mesozoic time this area was probably above the sea, so of course no marine sediments were deposited. (If any were deposited they were later eroded away completely—a picture geologists think is unlikely here.) In Cretaceous time, however, the sea once more swept across the Big Bend area, as well as across much of the rest of the North American continent. This time, the initial sediments were conglomerate and sandstone derived from nearby shores or islands and deposited in flat-lying layers in a probably shallow sea.

Later, as available sources for such coarse sediments were worn down, fine shale and limestone—the cliff-slope-cliff formations of Santa Elena Canyon—were deposited. Marine life was again abundant and varied, and shells of clams, oysters, and microscopic one-celled animals furnished much of the calcium carbonate of the limestone. Still later, thin layers of limy mud were deposited, now seen in the Boquillas Limestone near Hot Springs and in Boquillas Canyon. Marine reptiles, descended from land animals that returned to the sea, are known from these rocks, too.

Toward the end of Cretaceous time the sea once more receded. Sedimentation, however, continued, with delta, floodplain, and estuary sandstones and shales—the Javelina and Aguja Formations—being deposited over the older marine rocks. Today these strata yield fossilized trees and dinosaur bones, as well as remains of

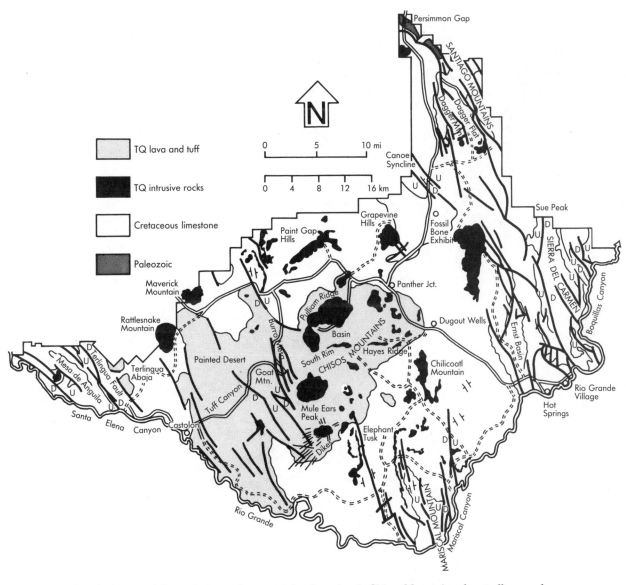

A geologic map of the park shows that except for the volcanic Chisos Mountains almost all exposed rocks are Cretaceous.

other land-type plants and animals. Perhaps the most remarkable fossil yet recovered from them is a giant flying reptile or pterodactyl with a 12-meter (50-foot) wingspan, by far the largest creature ever to soar the sky. The fossils all testify to a heavily vegetated tropical lowland supporting many plants and animals, similar perhaps to the present Amazon region. The sediments themselves came from newly formed but faraway mountains (as do the sediments of the Amazon region), for late Cretaceous and early Tertiary time saw the rise of the Rocky Mountains, birthplace of the Rio Grande.

Cenozoic Era. Uplift of the Rockies was followed, during Tertiary time, by erosion and by the depositing of vast quantities of mountain-derived gravel over the Great Plains and areas to the south and west. Development of the Basin and Range mountains did not come until 30 mil-

lion years later. As we've seen, most of these ranges are individual fault blocks lifted, tilted, even overturned in places, as if the whole region were a child's pile of blocks, their structures complicated by intrusions and volcanism. Here in the Big Bend region warping and fault movement occurred as the Sierra del Carmen and Mesa de Anguila blocks pushed upward, and as the central part of the park—the sunken block—sank. Within the sunken block, many smaller blocks were jumbled about as well, creating many of the lesser hills and mesas that now dot the desert surface.

These fault movements did not all take place at once. Rather, they occurred as many movements, a long series of earthquakes spread over millions of years, with displacement measuring in centimeters or, at the most, a few meters at a time. All in all, total movements along the main faults at

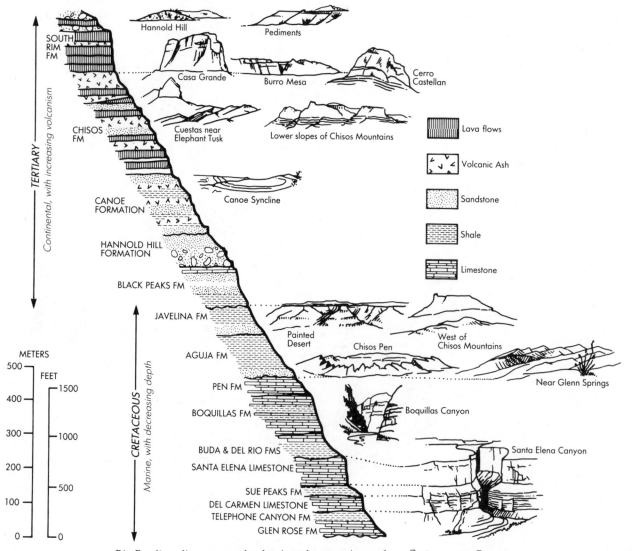

Big Bend's sedimentary and volcanic rocks range in age from Cretaceous to Recent.

the edges of the sunken block totaled nearly 1000 meters (3000 feet).

The Rio Grande's present canyons are about 500 meters (1500 feet) deep. All the upper layers of Cretaceous sedimentary rock—those now known only in the sunken block—have been stripped off Sierra del Carmen and Mesa de Anguila at the east and west ends of the park, providing good examples of the way in which uplift intensifies erosional effects. In Tertiary time tributaries blanketed low portions of the area with gravel, sand, and clay. These sediments, now partly hardened into rock, contain fossilized bones of many kinds of extinct mammals. Gone were the dinosaurs and the great swimming and flying reptiles, for the Age of Reptiles had ended and the Age of Mammals had begun.

Sometime during the Eocene Epoch, molten magma began to push up from reservoirs deep underground, particularly in an area near the center of the present national park. Wherever magma reached the surface, traveling up through fractures and fault zones, volcanoes spewed out great quantities of volcanic ash, and glowing lava flows spread across the land. Volcanic plugs, the hardened remains of magma in old volcanic conduits, tell us where some of these volcanoes were. In several parts of the park, magma unable to break through to the surface domed up over-lying Cretaceous sedimentary layers, forming laccoliths. Recent research suggests that some of the volcanoes may have collapsed, forming circular calderas.

By the end of Oligocene time, volcanism ended. Since then, except for the depositing of gravel and silt in late Tertiary and Quaternary times, the history of Big Bend has been one of erosion. Acting on a large, central mass of volcanic and intrusive rocks, erosion carved the varied topography of the Chisos Mountains. Stripping away much of the volcanic rock, particularly the easily eroded volcanic ash, which had barely consolidated into tuff, it left a few blocky caps of massive, resistant lava—the cliff-forming, mesa-topping rock of Casa Grande, Emory Peak, South Rim, and Burro Mesa. Erosion also exposed several pockets, large and small, of intrusive rock, as at Pulliam Ridge, Vernon Bailey Mountain, and the more isolated volcanic plugs and laccoliths around the park. Material eroded from the mountains adds further variety to the geologic picture: The terracelike gravel deposits seen in many roadcuts contain rounded cobbles and pebbles of volcanic and intrusive rock.

As you explore the park be sure to drive to Castolon and Santa Elena Canyon. There are geologic markers along the highway, which was routed to guide visitors past outstanding features of Big Bend geology.

At Boquillas Canyon the Rio Grande seems to vanish into a cleft in the limestone wall.

Burro Mesa Pouroff was once a giant pothole carved by a swirling whirlpool of water and rock fragments. The cave above the figures is deeply etched into a soft layer of tuff.

BEHIND THE SCENES

Boquillas Canyon. Big Bend's three major canyons all cut through uplifted Cretaceous sedimentary rocks. Where the Rio Grande enters Boquillas Canyon, the Santa Elena Limestone is faulted upward along the Sierra del Carmen fault zone; there it forms towering, nearly vertical cliffs. Downstream, beyond other faults, these rock layers are raised even higher, so that the canyon is walled with older Cretaceous rocks of the Del Carmen, Telephone Canyon, and Glen Rose Formations. Near the canyon the Santa Elena Limestone forms the crest of the Sierra del Carmen.

The sand slide near the canyon head is an unusual climbing dune blown against the cliff by down-canyon winds. The fine sand comes from the beaches, bars, and floodplain of the Rio Grande. Other dunes edge the river wherever the valley widens out.

Burro Mesa. The Burro Mesa tableland is topped with the same thick lava flow that caps South Rim and Emory Peak in the Chisos Mountains. On Burro Mesa the lava is about 1000 meters (3000 feet) lower than on those peaks, for here we are on the downdropped side of the Burro Mesa fault. Movement occurred along this fault after the lava was erupted and had cooled and solidified. Burro Mesa is now higher than its immediate surroundings because its hard lava cap resists erosion more effectively than surrounding layered sandstone and shale. Steeply bent rocks at the edge of the mesa were dragged into folds by movement along the fault.

Burro Mesa is shaped like a long, shallow celery dish. During heavy rains it drains at its south end through an unusual vertical channel known as Burro Mesa "Pouroff." The pouroff— not a geologic term but a very expressive name!—appears to have once been a deep, cylindrical pothole through which water that collected on the mesa surface spiralled to a lower opening, grinding its channel with boulders, pebbles, and sand. Rock fragments swirled in the pothole eventually pounded away one side of the vertical passage, opening the pothole to our view.

Casa Grande. A Chisos Mountain landmark, Casa Grande is an erosional remnant of the great pile of volcanic rocks that once covered this region. It is capped with massive lava flows of the South Rim Formation, here undermined by erosion of soft volcanic tuff. Where the resistant flows are undermined, slabs of lava break off along vertical joints and fall to the rocky talus-covered slope below.

Chisos Basin. Neither a volcanic crater nor a glacial cirque, the Chisos Basin has been mistaken for both. Encircled on the west and north by the resistant intrusive rocks of Ward Mountain and Pulliam Ridge, the basin drains through the Window, a narrow slot barely 3 meters (10 feet) wide (and a good place to keep a firm hand on your children). Apparently the hard intrusive rock on the west side of the mountains was less easily weathered and eroded than was the center of the mountain mass, which contains layers of volcanic tuff and even some sedimentary rock. Volcanic tuff and breccia on the south wall above Boulder Meadow have eroded into the spires and towers that give Pinnacles Trail its name.

Chisos Mountains. Near the center of Big Bend's broad central graben or sunken block, the Chisos Mountains are almost entirely volcanic. A single wedge of Cretaceous sedimentary rock runs through Laguna Meadow and Chisos Basin, a graben bounded on both sides by faults. Since the Cretaceous rock is more easily eroded than surrounding volcanic rock, it played a part in the shaping of these two depressions. Square-cut, cliff-edged Casa Grande, the South Rim, and Emory Peak—at 2570 meters (7835 feet) the park's highest summit—are all capped with the thick South Rim Lava Flow, once far more extensive than it is now. The same flow appears also on Burro Mesa, farther west and 1000 meters (3000 feet) lower because of movement on the Burro Mesa Fault. Faulting also played a strong part in raising the Chisos Mountains to their present height.

Rounded peaks such as Pulliam Ridge, Ward Mountain, and Vernon Bailey Peak are parts of an igneous intrusion, as are most of the outlying peaks of the Chisos Mountains. Lower slopes, ruled with horizontal or only gently tilted ledges, consist of Chisos Formation lava and ash flows— a case where volcanic rocks are stratified. These rocks are particularly evident on the west and south sides of the Chisos Mountains. Below them are sloping pediments of older rocks, Tertiary and Cretaceous sedimentary layers, cut in many places by dikes and volcanic necks. The

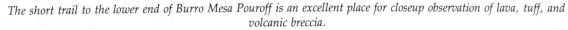

The short trail to the lower end of Burro Mesa Pouroff is an excellent place for closeup observation of lava, tuff, and volcanic breccia.

pediments merge outward with Late Tertiary and Quaternary gravel deposits.

Dagger Mountain. Probably a laccolith domed upward by an igneous intrusion, round-topped Dagger Mountain differs from other laccoliths in the park (Rattlesnake Mountain, Glenn Spring Laccolith, and Grapevine Hills among them) in that sedimentary rocks have not been worn away from its summit. The smoothly curving crest is surfaced with Santa Elena Limestone, the same resistant rock that caps Sierra del Carmen and Mesa de Anguila.

Elephant Tusk. This landmark is a volcanic neck, the hard infilling of a volcano's conduit. The magma that rose through it penetrated both the Cretaceous rocks that underlie the Chisos Range and the Tertiary Chisos Formation. Sloping ridges near the base of Elephant Tusk are

lava flows arching over the north end of a large anticline.

Fossil Bone Exhibit. Early Tertiary mammals buried in the sand and clay of a one-time streambed or floodplain are displayed in a roadside exhibit here. About 60 million years old, they occur in the Hannold Hill Formation, which is also exposed on the nearby ridge of rock. This is the youngest of Big Bend's prevolcanic sedimentary rocks; it overlies older and weaker layers of clay and sandstone of the Black Peaks Formation, which erode easily into gullies and badlands. Many of the newest-looking gullies may date from overgrazing that preceded establishment of the national park.

Grapevine Hills. Walk into the center of a laccolith! Molten magma, unable to escape to the surface here, domed up overlying sedimentary

On Pulliam Ridge, the north wall of Chisos Basin, weathering along finely spaced parallel joints has left bladelike ridges that stand out sharply from the mountain mass.

National Park Service photo.

*On the southwest flank of the Chisos Mountains, the Chisos Formation's
lava and ash beds are folded into a syncline.*

layers and then slowly cooled and hardened into the granite you see here. Overlying rocks have been stripped away, and weathering has attacked the granite along sets of parallel joints. Because sharp corners weather from two sides at once, blocks of granite become more and more rounded, a process known to geologists as spheroidal weathering. This type of slow weathering occurs both at and below the surface, with granite gradually turning into the coarse sand called grus. The non-rounded sand grains of grus are really the individual mineral crystals of the original granite. Grus usually contains both quartz and feldspar crystals. Since feldspars are softer and more brittle than quartz, and decompose into clay, most sand that has been transported any distance from its parent granite contains only quartz grains.

Hot Springs. Now contained in a rectangular manmade pool at the river's edge, these springs release nearly 85,000 liters (about 20,000 gallons) of hot water each day. Water temperature is about 40°C (105°F)—just right for a natural hot tub!

Several factors contribute to the presence of the hot springs:

• In this area the Earth's crust retains more than the usual amount of heat. The temperature increases more rapidly with depth than it does in most other, nonvolcanic regions.

• Groundwater is in plentiful supply, constantly replenished by the Rio Grande. However, because their flow is declining, many geologists feel that the springs derive their water flow from a "fossil" water source not replenished since perhaps Pleistocene time.

• Groundwater heated at depth tends to rise because of the temperature differential, just as water heated at the bottom of a kettle rises, causing boiling motions. Moving freely along some of

Mule Ears Peaks are two eroded dikes, harder than surrounding rocks and therefore standing up as erosional remnants.

the many joints and faults in this region, the heated groundwater reaches the surface rapidly, without substantial cooling.

• By carving deep into the land, the river has shortened the distance this heated water must rise before it emerges in springs.

Cliffs near the river are made up of layered marine sedimentary rocks, the Boquillas Limestone, a rock unit that is easily recognized by its light gray color and flagstone texture. Fossils of Cretaceous shellfish, large and small, are abundant in some of the limestone layers. (Please remember that fossils are protected by law in national park areas: no collecting!)

Lead-silver-zinc ores mined across the river in Mexico between 1890 and 1919 rode to a stateside smelter on an aerial cable tram that crossed the river downstream from present-day Rio Grande Village. Traces of the cable and its supports are still visible along the road to Boquillas Canyon. The 10-kilometer (6-mile) route had four cables, nine ore buckets, and 15 water buckets filled from the river.

Mariscal Mountain. Visible from the River Road or from the river itself, this mountain consists of a single large anticline sliced open by erosion. Almost a textbook example of an anticline breached along its summit, the fold as now exposed is centered with Santa Elena Limestone, and flanked with tilted younger Cretaceous rock layers. As seen in cross section, the anticline leans strongly westward and is broken along its axis by a single large thrust fault.

Mule Ears Peaks. These paired peaks east of Castolon are erosional remnants of an irregular, dikelike igneous mass that protruded through Tertiary rocks at the intersection of two faults. Much harder than adjacent rock layers, most of which have volcanic ash as a major component, they have remained above the surrounding landscape.

Painted Desert. North and south of Maverick (and in some other parts of the park as well) are heavily eroded areas of desolate, surrealistic beauty, where colorful layers of the Javelina For-

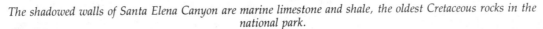

The shadowed walls of Santa Elena Canyon are marine limestone and shale, the oldest Cretaceous rocks in the national park.

Thick layers of tuff—once volcanic ash—form the walls of Tuff Canyon near the Castolon highway. A short trail leads down into the narrow gorge.

mation, a unit composed largely of volcanic ash, are exposed by erosion. Volcanic ash alters with time to a complex of clay minerals called bentonite. Because bentonite swells when it is wet and slakes easily when it is dry, it tends to weather and erode into badlands—as it has in

Badlands National Park, in the Painted Desert of Arizona, in parts of Yellowstone National Park, and in many other areas. In Big Bend's Painted Desert, fossil reptiles have been found in the pink, gray, yellow, black, and white tuff. Although the badlands existed long before the ar-

Not far from Tuff Canyon an interesting volcanic neck looks like an oversized petrified tree stump. The "stump" is 10 meters (30 feet) high and at the level of the "knothole" is 2.5 meters (8 feet) in diameter.

rival of the white man and his domestic animals, erosion here may have been aided and abetted by overgrazing.

Persimmon Gap. Hills on either side of Persimmon Gap contain remnants of an ancient folded and faulted mountain range, elevated a second time as part of the present Santiago Mountains. Slices of Paleozoic marine sedimentary rock are sandwiched between slabs of much younger rock—the Cretaceous Glen Rose Formation—in thin thrust-fault slices. The Gap itself is the abandoned channel of a former stream that cut across the southern Santiago Mountains in much the same manner that Dog Canyon, a few miles south, does today. Such a pass, no longer occupied by a stream, is known as a wind gap.

Rio Grande. Heading along the Continental Divide in Colorado, the Rio Grande flows south through southern Colorado and New Mexico, and then turns southeastward as the Texas-Mexico boundary. Where its course is deflected by desert ranges, it bends northeastward for a time in the "Big Bend" that gives this park its name. There it has incised three major clefts through the ranges: Santa Elena, Mariscal, and Boquillas Canyons.

In early Tertiary time, when the Rio Grande was just getting established, it probably flowed across a gradually sloping but fairly even surface. As basins and ranges of the southwest deserts developed, parts of its original course became impassable, and it was forced to wind its way among the ranges or to cut directly through them.

Santa Elena Canyon. Some 29 kilometers (18 miles) long, Santa Elena Canyon is carved almost entirely in the Cretaceous rocks shown in the stratigraphic diagram. Viewpoints east of Castolon are good places to match up the diagram and the rocks, starting with the Santa Elena Limestone forming the highest sheer cliff on both sides of the river, and ending with the Glen Rose Formation, partly covered with talus at the base of the cliffs. Younger (but still Cretaceous) rocks appear south of the canyon in Mexico, making up higher parts of Sierra Ponce. They are capped with a resistant sill of intrusive rock.

The narrow canyon, accessible by foot (in good weather) from the end of the road, is bordered with river-polished limestone. In places the gorge is only 60 meters (180 feet) wide; the walls are 500 meters (1500 feet) high, and as you can see they are in places nearly vertical. Think of the flood waters that used to roar through this nar-row opening before a long series of dams held back spring snowmelt from the Rockies!

Mesa de Anguila north of the river and Sierra Ponce south of it represent a single block lifted along Terlingua Fault, which lies along the base of the two mesas, crossing the river at the point where the Rio Grande emerges from Santa Elena Canyon. Both Terlingua Creek, entering from the north, and the Rio Grande follow the weakened rock along the fault zone.

Tuff Canyon. Just off the road to Castolon, Tuff Canyon is carved by Blue Creek into soft gray tuff, formed from volcanic ash. Blue Creek is dry most of the time. However it heads in the southern part of the Chisos Mountains, and is well supplied with water during and just after summer thunderstorms.

OTHER READING

Big Bend Natural History Association, 1980. *Road Guide to Paved and Improved Dirt Roads, Big Bend National Park.*

Big Bend Natural History Association, 1982. *Road Guide to Unimproved Backcountry Roads, Big Bend National Park.*

Maxwell, Ross A., 1968 (reprinted in later years). *The Big Bend of the Rio Grande.* Bureau of Economic Geology, University of Texas at Austin. (This book comes with a full-color geologic map of the park.)

Maxwell, Ross A., 1986. *Big Bend Country.* Big Bend Natural History Association.

Moss, Helen, 1984. "Big Bend: Where Mountains Seam the Sky." *National Parks*, vol. 58, nos. 11–12.

Capulin Mountain National Monument

Established: 1916

Size: 3 square kilometers (1.2 square miles)

Elevation: 2207 meters (7242 feet) at visitor center, 2494 meters (8182 feet) at highest point on crater rim

Address: Box 94, Capulin, New Mexico 88414

STAR FEATURES

• A shapely inactive volcano composed almost entirely of fine, lightweight volcanic cinders, with lava flows issuing from its base. A road leads to the edge of the crater, and a trail circles the rim.

• Views of many other volcanic features, including lava flows, lava dikes, cinder cones, shield volcanoes, and stratovolcanoes.

• An interpretive program that includes a film of the Mexican volcano Paricutin, similar in many ways to Capulin Mountain. Museum, trails, road to rim, and photographic roadside and trailside displays add to an understanding of the national monument.

SETTING THE STAGE

One of the many interesting volcanic features in northern New Mexico, Capulin Mountain is a cinder cone surrounded by small lava flows that emanated from its base. Its slopes are composed of volcanic cinders, pebble-sized pellets of light, bubbly volcanic material which were hurled skyward to fall to earth around the single volcanic vent. Near the vent, many of the cinders, still partly molten as they struck the ground, fused

The cindery slopes of Capulin Mountain rise about 300 meters (1000 feet) above the surrounding plain.

National Park Service photo.

An easy footpath encircles the crater rim, giving views of surrounding volcanic features.

with other fragments to form a loosely welded rock. Elsewhere, especially on the northeastern and eastern sides, the cinders lay loose and unwelded. Larger fragments thrown whirling into the air hardened into football-shaped volcanic bombs, many of which bounced down the cindery slopes to accumulate at the base of the growing cone. Because winds at the time of the eruption blew, as prevailing winds in this area do now, from the southwest, the airborne cinders tended to blow toward the northeast. As a result the crater rim is highest on its northeast side, lowest on the southwest.

Below the cone's western slope, jumbled blocks of several lava flows darken the surface. Their source is the *boca* (Spanish for "mouth") at the west base of the mountain. All the flows can be seen well from the Crater Rim Trail; those to the west are also apparent from the parking lot at the rim. Most of the flows north and south of the mountain consisted of thick, sticky lava that broke into jagged lumps as it moved. However a younger flow, curving around the base of Capulin Mountain, shows the distinctive surface features of quite fluid lava, with wavy pressure ridges arcing across it.

We know from the 1943 eruption of Paricutin Volcano in Mexico, and from studies of other recently active cinder cones such as Sunset Crater in Arizona, the many small volcanoes at Craters of the Moon in Idaho (both national monuments), and Cinder Cone in Lassen Volcanic National Park, that lava flows characterize the history of many cinder cones. Within underground magma chambers, dissolved magma gases tend to come out of solution and rise like froth when pressure is released during initial phases of an eruption, just as carbon dioxide bubbles come out of solution in a newly opened bottle of pop. As the froth escapes through the volcano's vent, pressure on the magma below is further lessened, and remaining gases expand even more rapidly. Frothy magma rushes up the escape route and shoots skyward explosively.

Other volcanic peaks and lava-capped mesas are seen from the route to the summit. There is no reason to believe that volcanism has ceased forever in this area.

National Park Service photo.

National Park Service photo.

Dry grass emphasizes the arcuate pressure ridges of a Capulin lava flow.

The tiniest dust-sized fragments are blown away by the wind; coarser ones fall back around the vent, accumulating to build up the cinder cone. Then, when the gaseous, frothy magma is expended, less gassy magma is likely to rise in the volcano's conduit, to escape as lava flows from the main crater or from pressure-produced fractures near the base of the cinder cone.

Capulin's lava flow basalt is dark gray, with many little rounded vesicles left by bubbles of gas. Its surface in many places has weathered yellowish or reddish brown because of oxidation of iron in the basalt. (The color is also enhanced by lichens growing on the rock.) Elsewhere, a thin coating of caliche (ca-LEE-chee) whitens the surfaces of lava blocks. If you look at some of this rock closely you will see that many vesicles are drawn out lengthwise by movement of still-molten lava.

The Capulin basalt is quite distinctive in its mineral composition. Tiny crystals of clear, shiny feldspar and of green olivine are scattered through much of it, though some is quite smooth-textured and lacking visible crystals, indicating especially rapid cooling. Lava ridges have piled up at the edges of some flows, much as ridges of wet leaves sometimes accumulate beside a gutter.

From the summit of Capulin Mountain you can see many other volcanic features: several other small cinder cones, some mesas capped with horizontal lava flows, and shield-shaped Sierra Grande to the southeast. Sierra Grande is composed mostly of lava flows, without ash or cinders, and as a result is less steep-sided than Capulin Mountain. It is also quite a bit older than Capulin, as we shall see. Differences in viscosity—runniness—of lava and its gas content determine whether a volcano develops as a cinder cone or as a shield volcano, or whether its flows flatten out horizontally leaving no perceptible mountain. Viscosity and gas content in turn depend on the chemical composition and temperature of the magma rising through the volcanic conduit. Frothy, gassy magma thrown upward explosively accumulates as cinder cones; less gassy lava, particularly that with the chemical composition of basalt, may flow quietly for considerable distances. Some flows in this area are covered with blocky, angular rubble, broken pieces of cooling lava crust carried along on top of still-molten material beneath.

GEOLOGIC HISTORY

Cenozoic Era. Capulin Mountain is one of the most recent of about 80 volcanoes in this area. All the volcanic features are Late Quaternary in age; they lie on top of Pliocene gravel and sand that elsewhere surface the High Plains. Volcanism was apparently intermittent from Pliocene to Holocene time, with at least three periods of volcanic activity alternating with periods of inactivity and erosion. The oldest signs of volcanism are the flat-lying lava flows known as the Raton Basalts, which cap long, high mesas on the northern skyline. In Pliocene and Pleistocene times the present scenery developed as erosion cut through these thick lava flows and carved out wide valleys in the soft sedimentary rock beneath. These valleys then became the sites of younger volcanic features.

Lavas of the second phase, known as Clayton Basalts, occur under and around Capulin Mountain. They form Emery Peak northeast of Capulin and Sierra Grande to the southeast, as well as Mud Hill and almost all the other mesas and hills visible from Capulin's rim. Now modified by erosion, most of the flows have smooth, grassy surfaces with only occasional hillocks of bare basalt. Some of the Clayton flows temporarily dammed the Cimarron River, creating a lake; others later replaced the lake water with lava.

Capulin Mountain itself, as well as nearby Baby Capulin and Twin Mountain, erupted during the third and last volcanic stage. Already on the scene here was Folsom Man, who 10,000 years ago prowled this area for game, hunting with stone-tipped spears. Folsom Man gets his name from the town of Folsom, just northeast of Capulin, where his carefully shaped spear points were found associated with bones of extinct bison. Some of Folsom Man's descendants camped on a tongue of Capulin lava, leaving charcoal that can be dated, bracketing the birth of the volcano between 8000 and 2500 B.C. So these early people, eyes wide with trepidation, may well have fled the "supernatural" fireworks of Capulin and its small neighbors!

Little erosion has occurred since these eruptions, except at the hand of modern man, who seeks volcanic cinders for railroad ballast and cinder blocks. Twin Mountain and Baby Capulin have been quarried for this material; Capulin has been spared, thanks to National Park Service protection. Capulin's blanket of cinder supports

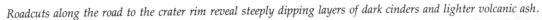

Roadcuts along the road to the crater rim reveal steeply dipping layers of dark cinders and lighter volcanic ash.

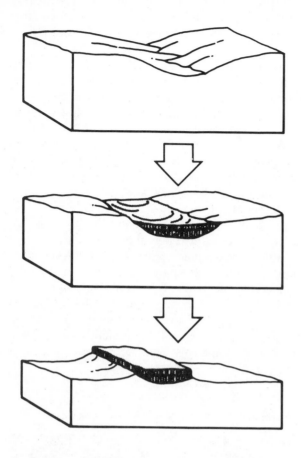

Some of the lava in the Capulin area flowed down pre-existing valleys without covering bordering ridges. When erosion wore away the unprotected ridges, the lava-capped valleys remained as narrow, sinuous mesas—a reversal of the original topography.

a growth of pinyon pine, juniper, scrub oak, mountain mahogany, and various other plants whose roots protect it from rapid erosion. Several neighboring cinder cones unprotected by such growth are now mere rounded hills.

OTHER READING

Christiansen, P. W., and Kottlowski, F. E. (editors), 1972. *Mosaic of New Mexico's Scenery, Rocks, and History.* Scenic Trips to the Geologic Past, no. 8, New Mexico Bureau of Mines and Mineral Resources.

Muehlberger, W. R., Baldwin, B., and Foster, R. W., 1967. *High Plains, Northeastern New Mexico-Raton-Capulin Mountain-Clayton.* Scenic Trips to the Geologic Past, no. 7, New Mexico Bureau of Mines and Mineral Resources.

Carlsbad Caverns National Park

Established: 1923 as a national monument, 1930 as a national park
Size: 189 square kilometers (73 square miles)
Elevation: 1340 meters (4400 feet) at visitor center
Address: Box 1598, Carlsbad, New Mexico 88220

STAR FEATURES

• A radio-guided stroll through the subterranean world of Carlsbad Cavern, one of the largest known solution caverns in the world. Profusely adorned with stalactites and stalagmites, the cavern also displays many columns, draperies, cave pearls, and other unusual deposits.
• New Cave, an undeveloped cavern (no lights, no trails, no elevators) that can be visited on guided tours only.
• Part of a 250-million-year-old reef and related lagoon and shallow marine deposits.

SETTING THE STAGE

Limestone caverns nearly always form by a two-cycle process of dissolving, which forms the cavern system, and depositing, which ornaments it. Carlsbad Caverns are no exception. Here geologic and climatic factors have combined to create a labyrinth of unusually large chambers and corridors, with many interconnecting passageways, an intricate cavern system of great beauty, with ornate and varied adornment.

Groundwater is the chief agent of cave solution, a process that is largely chemical in nature. Rain or snow that falls through air and filters

National Park Service photo.

In the Hall of Giants, frail stalactites and fine soda straws contrast with massive stalagmites and rippled onyx draperies. Stalactites, spelled with a c, hang from the ceiling; stalagmites, spelled with a g, rise from the ground.

through soil becomes mildly acid. The acid is carbonic acid, resulting from combining carbon dioxide with water. As slightly acid water seeps into limestone, especially near the water table (the surface below which the rock is completely saturated), it little by little dissolves and carries away some of the calcium carbonate of the limestone. Above the water table, percolating groundwater tends to deposit calcium carbonate as well as to dissolve it. Trickling or dripping through air-filled spaces, the water leaves behind some of its calcium carbonate, usually deposited as travertine. Flowing water deposits flowstone, dripping water deposits dripstone—both the same in composition but forming different types of cave ornaments.

Recently another mechanism was suggested that may have helped in cave-forming processes. The chemical reactions outlined above involve solution of limestone by carbonic acid. Solution by sulfuric acid may also have occurred, with sulfur derived from the mineral pyrite, known to have been present in some of the lagoon limestones in which the caverns are located. Instead of simply dissolving limestone, sulfuric acid combines with it to form calcium sulfate, the mineral gypsum. (Gypsum may also have been brought in from overlying Permian and Triassic rocks, which contain a good deal of it.) Later, the gypsum might decompose in the sulfur-enriched air of the cavern, in which case the cave would gradually hollow out. Since the process could take place in an air-filled cavern, it would not require that the cave be below the water table during its enlargement. This process would also explain the presence of several very large blocks of gypsum and a scattering of gypsum ornaments in some parts of the cavern.

Stalactites and stalagmites of the Big Room formed along joints where groundwater penetrated the limestone. They are composed of almost pure, finely crystalline calcite.

Gently dipping layers of reef limestone guide runoff and groundwater toward the edge of the reef. This view looks southwest along the curving reef escarpment. The visitor center and cavern entrance are below the arrow.

National Park Service photo.

GEOLOGIC HISTORY

Paleozoic Era. The limestone strata of the Guadalupe Mountains, in which Carlsbad and at least 50 other caves have formed, had their beginning in the Permian Period, about 250 million years ago. Most of the limestone accumulated near the edge of an ancient bay as the long Capitan Reef, which outwardly resembled modern coral reefs such as Australia's Great Barrier Reef. Development of the thick limestone layers of this region is discussed and illustrated in the chapter on Guadalupe Mountains National Park.

Mesozoic Era. Reef growth eventually stopped, probably because increasing saltiness of water in the nearly landlocked sea killed the reef organisms. Not long after it ceased growing, the massive reef began to settle over the loosely packed, broken reef material underneath it. At the same time it began to crack. The cracked or jointed reef rock, formed from the skeletons of myriad lime-secreting algae and sponges, can be seen at many places along the entrance road as well as near the cavern entrance. Most of its joints are vertical, and many are arranged in parallel sets. Prominent sets trend either in the same direction as the reef escarpment, about northeast, or perpendicular to the escarpment, southeast. These joints and their two-direction arrangement are responsible for the floorplan of the cavern. At the

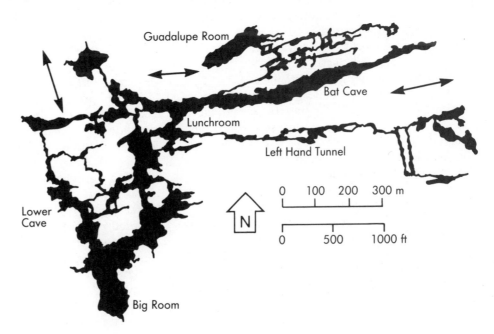

Guadalupe Room

Bat Cave

Lunchroom

Left Hand Tunnel

Lower Cave

Big Room

0 100 200 300 m

N

0 500 1000 ft

If you could look down on Carlsbad Cavern with X-ray eyes, you would see several long, fairly straight passages where solution followed major joints.

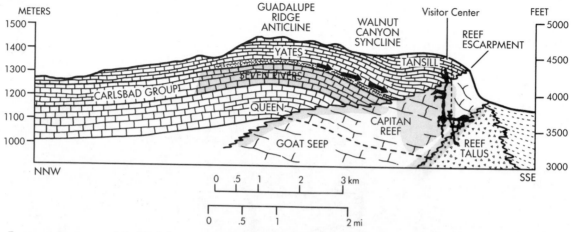

Rooms and passages of Carlsbad Cavern are dissolved in limestone of Capitan Reef. Sloping lagoon deposits shunted rainwater toward the reef escarpment, as shown by arrows. The four shaded areas are major solution zones. (Adapted from D. H. Jagnow.)

surface you can see that rainwater and small rain-fed rivulets widen the joints; underground, water seeping through the joints does the same thing. As might be expected, many caverns in the Guadalupe Mountains have passages that follow joints, some for hundreds of meters.

Cenozoic Era. The caverns' beginnings can be traced to uplift of the Guadalupe Mountains, which started some 12 million years ago. Uplift initiated cavern development, allowing groundwater to move through the limestone. Upward movement was greatest along the fault zone that edges the west side of the range. Hinged along the east edge, the mountains simply tipped up at an angle. They arched a little, too, so that in the Carlsbad area the summit of the arch forms Guadalupe Ridge, the skyline northwest of the visitor center. Parts of the range now tower 1200 meters (4000 feet) above the surrounding desert. Near Carlsbad Caverns the relief is not so great—about 450 meters (1500 feet).

The steep slope along the east side of the range is the seaward edge of the Capitan Reef, the reef escarpment, preserved for nearly 250 million years. It is prominent now because erosion has removed most of the soft deposits of the former bay, which lay east of the escarpment and received such easily eroded or even completely soluble minerals as salt and gypsum, as well as lots of soft, silty sandstone and limestone.

Some parts of the reef, especially its contacts with the lagoon deposits above it and the reef talus below it, were more easily dissolved than others, and so were more likely to be the sites of cave development. Relatively impermeable silt or clay layers in the lagoon deposits influenced the location of the caverns, too, for they shunted

groundwater toward the front of the reef, where the caverns are located.

The water table also controlled the shape of the caves, for it determined the depth at which acidified water most actively dissolved the limestone. The water table in turn reflected the amount of uplift of the mountains. Here, uplift was an installment-plan affair lasting from Miocene time through the Pliocene and Pleistocene Epochs and even into Holocene time—a span of 12 million years. The highest cave levels formed first, when the mountains were not nearly as high as now. Then as the range continued to rise the water table stabilized at lower and lower levels. So the highest cave levels were lifted, drained, and slightly tilted, while new cave levels began to develop below them. Every prominent level represents a long period when the water table remained relatively stationary. There is some evidence that the caverns once included levels higher than those we know today: Flowstone and dripstone blocks have been found on the surface above the present cavern system.

Other caves in the Guadalupe Mountains are similarly controlled by geologic features such as jointing, solubility of certain parts of the reef, surface drainage, periodic uplift, and corresponding periodic changes in the water table. For all we know, new chambers are forming at the surface of the present water table, far below frequented parts of these caverns.

Once the many interconnecting channelways formed—whether by carbonic or sulfuric acid—nonchemical processes enlarged them. Thin partitions and weak ceilings must certainly have collapsed, especially when support-giving water drained away. (Be reassured: No major collapse

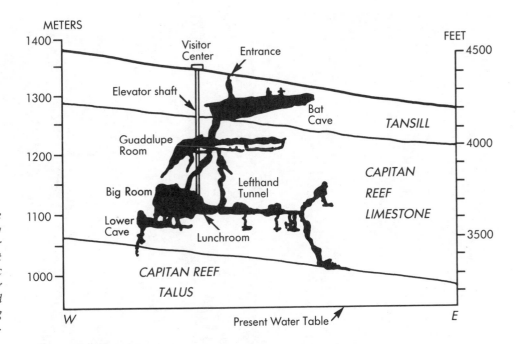

METERS

1400

1300

1200

1100

1000

FEET

4500

4000

3500

Visitor
Center

Entrance

Elevator shaft

Bat
Cave

TANSILL

Guadalupe
Room

Big Room

Lefthand
Tunnel

CAPITAN

REEF

LIMESTONE

Lower
Cave

Lunchroom

CAPITAN REEF

TALUS

W

Present Water Table

E

Carlsbad's distinct cave levels were dictated by position of the water table, and show how it dropped with periodic uplift of the reef. Upper cave levels are tilted by continuing hingelike uplift.

National Park Service photo.

Some rippling draperies are so thin you can see light through them. Their irregular upper surfaces tell of splashing water drops that trickled over the edge to form the draperies.

Stalactites and stalagmites grow by tiny additions of calcium carbonate. Eventually they join to form columns.
Popcorn was added during a later resubmergence.

has occurred since the cave's discovery more than a century ago.) As you stroll through the cavern, look for examples of exfoliation, where limestone roof and walls seem to crack away in large flat or curving slabs. Exfoliation (also called spalling) is really another type of jointing, with joints parallel to exposed surfaces. Commonly it is caused by removal of pressure from the face of the rock. When rock is dissolved away, pressure on the remaining "parent" rock is relaxed, and it expands just enough to crack away sheets of weathered rock from its surface. Exfoliation slabs range in thickness from a few millimeters to half a meter or more (fractions of an inch to 2 feet). Each time a slab falls, the cave is higher or wider. The extra weight of stalactites and other ornaments helps to pull away loosened slabs. Exfoliation occurs at the surface, too; there are many examples of it in Walnut Canyon along the park entrance road.

Fairly late in the history of the caverns, after cavern rooms and passageways were lifted above the water table, second-stage processes began—the formation of dripstone and flowstone ornaments. The ornaments seem to be quite young, in geologic terms—Pleistocene and Holocene— but ornamentation of upper levels may have begun while lower levels were still being opened up. With every uplift, groundwater drained away, and more of the cave was dry enough to become decorated. Surface water, still sinking into the ground, still becoming acid, absorbed limestone and carried it downward to the cavern chambers and passages. As each drop trembled on the ceiling or flowed thinly down a wall, it deposited its tiny load of calcium carbonate as travertine. Drop by drop, stalactites formed, lengthening and thickening through thousands of years. Intricate waterfalls and sinuous draperies, splatter-topped, many-tiered stalagmites, and fine translucent sheets of flowstone all tell the same story.

Innumerable variations in cave decorations, with their mysterious shapes and shadows, give variety to the tale of the past. Stone lily pads and cave popcorn adorn the edges of quiet pools of

Scalloped lily pads develop when water levels remain stable for a time. Calcium carbonate concentrates near the water surface, where carbon dioxide escapes into the air.

National Park Service photo.

water; in some cases they tell of long-gone pond levels. Partly dissolved stalactites and stalagmites or horizontal lines of popcorn on the chamber walls tell of a rising water table, when portions of the cave were resubmerged or temporarily flooded. Gravel deposits on the cavern floor, some of them covered with coats of flowstone, say that streams once flowed through parts of the cave—perhaps during the rainy cycles that accompanied the Ice Ages. Unbroken stalactites on a huge fallen block affirm that it fell gently into a water-filled chamber, and that the

National Park Service photo.

Bizarre popcorn ornaments that add to the gothic grandeur of the Kings Palace indicate a second submergence.

The cavern entrance is framed with Permian limestone deposited behind Capitan Reef. The entrance is a natural one, created by collapse of part of the cavern roof.

Gnarled, fingerlike helictites seem to defy gravity. Center openings of growing helictites are so small that capillary action and fluid pressure become more important than gravity in controlling the direction of their growth.

water cushioned the fall so well that delicate dripstone ornaments remained intact. As the climate became drier, the water table dropped once more, leaving the caverns as they are today.

BEHIND THE SCENES

Carlsbad Cavern. By all means take one of the radio-guided walks through this celebrated cavern. Choose the walk you can handle, go at your

Many dripstone features of New Cave match those of Carlsbad Cavern. The "Christmas tree" in New Cave shows snowlike deposits on top of a highly ornamented stalagmite.

National Park Service photo.

own pace, stop often just to look. You'll need a sweater, as the cavern stays at a fairly steady 13° C (56° F).

The entrance to the cavern is surrounded by flat-lying layers of limestone deposited in the Permian lagoon behind Capitan Reef. The entrance is a natural one, a sinkhole created by collapse of part of the cavern roof. The pathway descends steeply into the reef limestone, and flattens out along levels predetermined by later water table levels. The best views of the limestone are outside, for inside the cavern the walls, ceiling, and floor are nearly everywhere covered with travertine in its many ornamental forms. The cavern's rooms and passages penetrate an area 1500 meters long, 900 meters wide, and 300 meters from top to bottom (5000 × 3000 × 1000 feet).

New Cave. Many of the dripstone features of New Cave match or surpass those of Carlsbad Cavern. Visits are with Park Service guides only, and are somewhat strenuous, as the walk to the cave entrance is uphill and no trails have been built in the cave itself. You'll need good walking shoes, an old sweater, and a reliable flashlight. Most of the features described above for Carlsbad Cavern apply also to New Cave, though this cavern does not seem to be as strongly controlled by ancient water table levels.

OTHER READING

Barnett, John, no date. *Carlsbad Caverns National Park.* Carlsbad Caverns Natural History Association.

Christiansen, P. W., and Kottlowski, F. E. (editors), 1972. *Mosaic of New Mexico's Scenery, Rocks, and History.* Scenic Trips to the Geologic Past, no. 8, pp. 76–78. New Mexico Bureau of Mines and Mineral Resources.

Hayes, P. T., 1957. *Geology of the Carlsbad Caverns East Quadrangle, New Mexico.* U.S. Geological Survey map GQ 98, with text.

Hayes, P. T., and Koogle, R. L., 1958 (reprinted 1975). *Geology of the Carlsbad Caverns West Quadrangle, New Mexico–Texas.* U.S. Geological Survey map GQ 112, with text.

New Mexico Geological Society, 1954. *Road Log, Guadalupe Mountains area, New Mexico and Texas.* New Mexico Geological Society Guidebook, 5th Field Conference, pp. 40–64.

Chiricahua National Monument

Established: 1924
Size: about 50 square kilometers (19 square miles)
Elevation: 1562 to 2385 meters (5124 to 7825 feet)
Address: Dos Cabezas Route, Box 6500, Willcox, Arizona 85643

STAR FEATURES

• Storied towers arrayed against the mountain, with balanced rocks and tall narrow alleys in layers of volcanic tuff. The tuff tells a tale of explosive eruptions from a large volcano southeast of the present national monument.

• Excellent and interesting trails, some with descriptive leaflets.

• Views of Cochise Head to the northeast, and of Willcox Playa, relic of an Ice Age lake, to the northwest.

• An interpretive program that includes museum exhibits, nature walks, and evening programs.

SETTING THE STAGE

Cardboard armies seem at first glance to march in grand array above the canyons and mountain slopes of the northern Chiricahua Mountains. Just what are these strange pink statues, and why do they erode into such bizarre figures? The rocks themselves are rhyolite tuff, fine-grained volcanic ash born in the maelstrom of huge eruptive explosions that burst from craters some distance southeast of the present national monument. Molten magma from an underground magma chamber, seething with steam and other rapidly expanding gases, increased vastly in volume as it emerged. Much of it was no doubt carried skyward in great roiling gray clouds like those put forth by Mount St. Helens in 1980, but other ash clouds swept furiously downslope as swift avalanching flows of fine, glowing particles. When these particles settled, still incandescent, they were so hot that they fused or welded together into rock known as ash-flow tuff (in contrast to ash-fall tuff, which falls from ash clouds and is cool by the time it reaches the ground). Here, the varying degrees of welding give us clues to some of the horizontal banding and corrugation seen in the eroded figures, for strongly welded ash alternates with much less strongly welded or even nonwelded material.

The sequence of these tuffs has been carefully studied. As each ash flow came to rest, the thin layer that first touched the ground cooled so quickly that it did not everywhere fuse, but re-

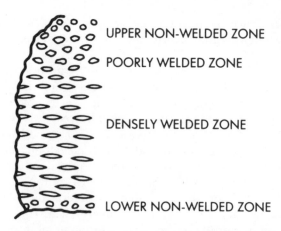

UPPER NON-WELDED ZONE
POORLY WELDED ZONE

DENSELY WELDED ZONE

LOWER NON-WELDED ZONE

A single cooling unit of volcanic tuff contains a densely welded layer sandwiched between thinner, less strongly welded zones. Weathering characteristics and the shape of inclusions are clues to the three layers of each cooling unit.

Tad Nichols photo.

Apache warrior Cochise sleeps beyond a forest of rock pinnacles eroded in tuff from a volcanic extravaganza.

mained as a thin nonwelded zone. Immediately above this initial layer, and insulated from the cool ground surface by it, the hot volcanic ash cooled much more slowly. Compressed and kept hot by additions of yet more glowing ash, it formed a welded zone in which glassy ash fragments coalesced completely into rock that has many of the characteristics of stoneware or porcelain. The ash-flow tuff will, for instance, "ring" when struck with a hammer. (Take my word for this. The Park Service frowns on rockhammers!) Densely welded zones grade upward into less compressed but still partly welded overlying material, and finally into other zones of nonwelded tuff that cooled rapidly in contact with air.

You can identify some of these zones without much difficulty along the Echo Canyon Loop Trail. In both the upper and lower nonwelded zones, light gray inclusions in the darker tuff are almost equidimensional as seen from the side or from the top. Inclusions in the partly welded zones, on the other hand, are slightly squashed. Those in the densely welded zones are flat as coins, and when seen edge-on appear as thin streaks. Some of the densely welded tuff is quite red. Expect to see these variations repeated, for the trail climbs and descends through six superimposed ash flows.

By and large, weathering by water and wind follows joints in the rock, particularly the promi-

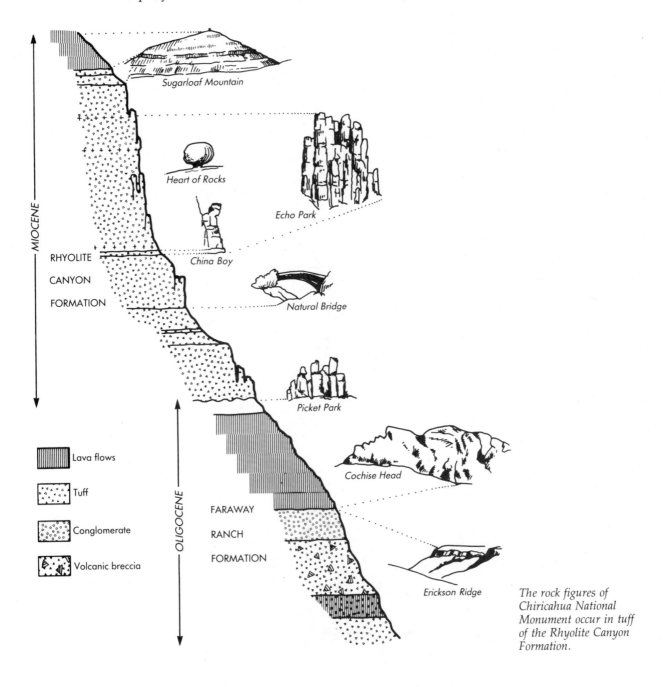

MIOCENE

RHYOLITE

CANYON

FORMATION

Sugarloaf Mountain

Heart of Rocks

Echo Park

China Boy

Natural Bridge

Picket Park

Cochise Head

OLIGOCENE

FARAWAY

RANCH

FORMATION

Erickson Ridge

Lava flows

Tuff

Conglomerate

Volcanic breccia

The rock figures of Chiricahua National Monument occur in tuff of the Rhyolite Canyon Formation.

nent sets of vertical fractures that result from earth movement along parallel northwest-to-southeast faults. Some of the offset on these faults is horizontal, one side sliding northwest past the other. As the fault blocks moved, the rocks broke along three sets of parallel joints, as shown in the diagram. The northwest-trending and northeast-trending vertical joints can be distinguished easily just about everywhere in the national monument. East-west-trending joints are not as well defined. Here and there, nonvertical joints show up as diagonal slashes. Along the trails, where wind has swept them clear, you can often pinpoint the joints and see that alleys between pinnacles follow them exactly. On the east

slope of the mountains, visible from the road to Massai Point, vertical fractures are not as well developed, and pinnacle armies do not appear.

As water and wind eroded the fractured rock, the joints little by little were widened. The task was easiest in the chalky rock of unwelded zones, more difficult in partly welded zones, and most difficult in densely welded zones. The relative softness or hardness encountered within the zones brought about the horizontal corrugations you see on the pinnacles and other figures.

Balanced rocks are unusually common here. There is some disagreement, though, on how they form. All geologists do agree that weathering of hard and soft rock layers initiates the

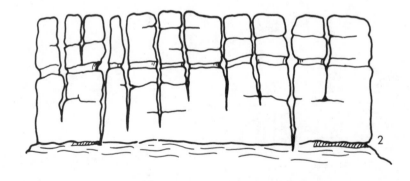

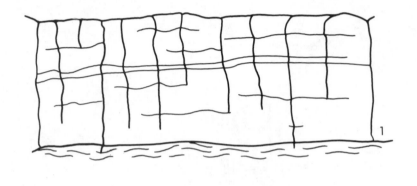

The storied pinnacles developed as welded tuff eroded along vertical joints and horizontal weak zones.

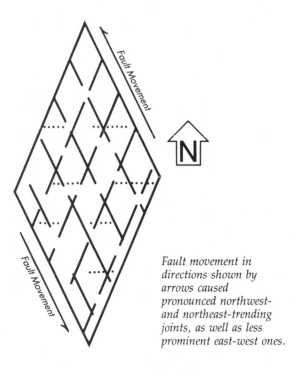

Fault movement in directions shown by arrows caused pronounced northwest- and northeast-trending joints, as well as less prominent east-west ones.

process, undermining blocks of harder, more resistant rock by wearing away the softer, less resistant rock below. A logical theory for their further development surmises that as the top of an undermined pedestal begins to tip toward one side, the softer rock below that side becomes compressed—and therefore more resistant—on that side. The compression gives erosion a chance to work away at the other side until the block on top begins to lean slighty in *that* direction, again speeding up erosion on the unburdened side. As wind and water wear away first one side and then another, the "stem" becomes narrower and narrower. Eventually, of course, the balanced rock will be so precariously perched that it will tumble off its pedestal. But the national monument won't be short of balanced rocks, as others are continually forming.

Several other rock types appear in the monument. Capping Sugarloaf Mountain is a layer of dark gray, white-spotted rhyolite that is similar

Big Balanced Rock is one of many balanced rocks in this monument.

in composition to the tuff layers, but that formed as a lava flow. Along the Bonita Canyon road are some flashy red sedimentary rocks—conglomerate, sandstone, and siltstone—that are thought to have been deposited in a small, short-lived lake, possibly even a hot one, dammed by lava flows. White stringers threading through these rocks are gypsum veinlets.

Another interesting rock, visible near and downstream from the visitor center, is a bouldery conglomerate made of volcanic rocks reworked by streams. It was formed about 30 million years ago, before the eruption of the volcanic ash of the monument's pinnacles, and it closely matches in appearance the gravel deposits of alluvial fans forming today on slopes around the present desert ranges.

In a small area along the Echo Canyon Loop Trail the ground is sprinkled with small, gray, marble-sized balls that look somewhat like hailstones. They *are* hailstones of a kind, for they develop, as do ice hailstones, in the violent updrafts within turbulent clouds. In the great cauliflower-shaped clouds of volcanic eruptions, hot particles of ash and pumice are tossed about, enlarging by collision with other ash and pumice particles. Volcanic hailstones have been observed also during modern eruptions.

Chiricahua is a great place to look at lichens; they seem to find rhyolite tuff particularly attractive. These algae-fungi communities initiate the processes that break rocks down into soil. Other plants, and animals as well, also influence geologic processes here by excreting acid wastes that dissolve cementing material between rock grains. Plants have purely mechanical effects, too, as their roots penetrate and grow in narrow crevices.

Colorful lichens, pioneers of the plant world, decorate trailside tuff.

GEOLOGIC HISTORY

Precambrian, Paleozoic, and Mesozoic Eras. The earliest part of the history of the Chiricahua area is not recorded within the national monument. The road from Willcox to the monument, however, crosses some Precambrian granite and passes through a prominent hogback, a ridge of tilted, hard, pebbly Cambrian quartzite. Younger Paleozoic and Mesozoic rocks occur on the margins of the Dos Cabezas and Chiricahua Mountains, so we can piece together a history that is fairly typical of southwestern United States: predominantly marine sedimentation in Paleozoic time, predominantly nonmarine sedimentation and volcanism during the Mesozoic Era.

As the Mesozoic Era drew toward its close, volcanism continued to mark this region. Lava and ash flows spread outward from volcanic centers, creating a large volcanic field that included not only the locale of the present Chiricahua Mountains but a wide area extending south and southeast into Mexico. The volcanism began about 120 million years ago and lasted 100 million years.

Cenozoic Era. The oldest volcanic rocks now exposed within Chiricahua National Monument formed less than 40 million years ago. These rocks occur only in the low western parts of the monument and in and near Cochise Head. Since the volcanic layers thicken markedly toward the southeast, their source is thought to be in that direction. Lava and ash flows, breccia made up of angular volcanic fragments, and the alluvial fan gravels described above date from these eruptions. The rocks are fairly rich in iron minerals, more so than younger volcanic rocks that make up Chiricahua's erosional armies, and as a result they are darker in color and may be classed as rhyodacite, a rock intermediate between rhyolite and dacite.

A long period of erosion followed the series of eruptions that created these rocks. Then about 25 million years ago eruptions began again, this time from a volcanic center a few miles south of the present national monument. One by one, six major ash flows and a final lava flow burst from this center, streaming out through cracks and fissures caused by upwelling of magma. Violent explosions repeatedly shot out the devastating incandescent ash flows that became today's layers of ash-flow tuff. As magma gases were exhausted, the eruptions became less violent; molten magma rose in the volcano's conduit, and a sheet of lava poured out over the ash layers.

Finally the roof of the partly drained magma chamber, no longer supported from below, collapsed, creating a large, elliptical caldera, and the eruptions came to an end.

In the last 10 million years, block faulting lifted the Chiricahua Mountains and the surrounding ranges relative to intervening basins. Coupled with faulting came the joints that set the stage for erosion of the monument's pinnacles and spires. Streams stripped away many hundreds of feet of surface rock, carrying boulders and cobbles, sand and mud into nearby valleys, reducing the height of the mountains and concealing their real bases with valley fill.

In Pleistocene time, while most of North America and Eurasia shivered in the grip of Ice Age glaciers, this area also underwent cyclic climate extremes in the form of alternately wet and dry cycles. During rainy parts of the cycles, enclosed basins between the ranges held lakes, some of them freshwater, others salty. The mark of one such lake can be seen from high points within the national monument: Willcox Playa to the northwest, white-floored with precipitated salts. The lake that once existed here, known to geologists as Lake Cochise, may never have risen high enough to overflow southward into Mexico. But it covered a large part of Sulphur Springs Valley north and south of Willcox. All these bodies of water dried up as the rainy epochs ended about 8000 years ago. After heavy rains, Willcox Playa may be flooded, forming a new and very temporary "lake" only a few inches deep.

OTHER READING

Jackson, Earl, 1970. *The Natural History Story of Chiricahua National Monument*. Southwest Parks and Monuments Association (out of print but available in libraries).

Marjaniemi, D. K., 1968. *Tertiary Volcanism in the Northern Chiricahua Mountains, Cochise County, Arizona*. Arizona Geological Society, Southern Arizona Guidebook III.

Southwest Parks and Monuments Association, no date. *A Guide: Mountain Island Environmental Study Area, Chiricahua National Monument*.

Yetman, R. P., 1974. *The Geology of Chiricahua National Monument*. Mimeographed; request at visitor center.

Death Valley National Monument

Established: 1933
Size: 7770 square kilometers (3000 square miles)
Elevation: 86 meters (282 feet) below to 3368 meters
  (11,049 feet) above sea level
Address: Death Valley, California 92328

STAR FEATURES

• The lowest place in North America, a below-sea-level basin whose bedrock floor dropped thousands of meters between elevated fault-block ranges.

• Pine-clad mountains, snowcapped in winter, within sight of the searing valley. Telescope Peak supports a stand of aged bristlecone pines.

• Barren mountainsides inscribed with complicated, hard-to-decipher geologic patterns caused by faulting, folding, and super-sized landslides.

• Desert playas, salt pans, and old beach ridges that testify to lakes that once occupied Death Valley's now arid depression. Interesting salt deposits can be seen close at hand.

• Borax deposits resulting from hot spring activity.

• Other desert features: sand dunes, badlands, giant alluvial fans, desert pavement and varnish, narrow rockbound canyons telling of centuries of drought and flood, and unusual rocks that move with the wind.

• Explosion craters and other volcanic features stemming from the rise of molten rock along fault zones.

• Visitor center, museums, scenic drives and trails (many with guide leaflets). The interpretive program includes talks and guided tours.

See color pages for additional photographs.

SETTING THE STAGE

Why are we fascinated by rocks pushed by the wind? By the lowest spot on the continent? By steam explosion craters and evidence of their violence? By hotness, salt, and the strangers who named this great depression "Death Valley?"

This flat-floored, zigzag trough of a valley is the lowest, hottest, and driest mountain-fringed basin of the desert southwest. A graben, it is a product of fault movements rather than of stream or river excavation. (Where would the rivers *go*?) Subsidence between two major faults has put parts of its floor 86 meters (282 feet) below sea level.

Ranges bordering Death Valley—the Panamint Range and Cottonwood Mountains on the west and the Grapevine, Funeral, and Black Mountains on the east—are edged with the still-active faults that shaped, and continue to shape, the valley. Time-scarred, etched by gravity, wind, and infrequent but often torrential rain, the mountain faces rise above spectacular alluvial fans produced where steep mountain canyons disgorge into the valley. The fans represent in only a minor way the vast amount of debris actually removed from these mountains. Gravel deposits of old alluvial fans, along with lake deposits, landslides, and accumulations of volcanic ash, are more than 3000 meters (10,000 feet) thick beneath the present fans and salt pans. Reconstructing the original mountains and adding their height to the depth of these sediments, we find that total vertical movement on the valley-edge faults has been at least 6700 meters (about 4 miles).

As seen from Dantes View, the salt-glazed floor of Death Valley stretches north between parallel ranges. Offset by cross-faults, the valley zigzags to left, then right beyond the distant hills.

Recent movement along one of the Black Mountain faults has created a visible step (short arrow) in the profile of an alluvial fan near Badwater. An older step (long arrow) marks an earlier break.

Important factors in creating the Death Valley of today, mountain uplift and valley downdropping, have kept well ahead of erosion and infilling. Death Valley has no outlet. Streamflow from the surrounding mountains, some of it relayed by adjacent higher-elevation valleys that in turn drain into Death Valley, simply dries up, evaporates, leaving its dissolved minerals behind. Rainfall here in the rainshadow of the Sierra Nevada averages no more than 4.3 centimeters (1.7 inches) per year; some years see no rain at all. Given a chance, the hot, dry valley air could evaporate as much as 3.8 *meters* (12.5 *feet*) of water each year. With more rainfall in the surrounding mountains, and with considerably less evaporation within the valley, Death Valley would contain a long, narrow, probably salty lake, as it has more than once in the past, particularly when it was fed also with glacial runoff from the Sierra Nevada to the west.

Normal faults, caused by tension or pulling apart in the Earth's crust, are responsible for most of the valley's depth. However horizontal movement on some faults has played an important part in opening up Death Valley, too. In keeping with the general pattern of many major faults in California (including the San Andreas Fault), land on the west side of these faults has moved northward or northwestward relative to land east of the faults.

Death Valley is quite young in a geologic sense. Its bordering faults are young. The straight-fronted ranges that rise above its floor are young, as is shown by their steep, craggy slopes and faceted ridges, by shattered rock debris still clinging to their crags, and by the "wineglass" shape of some of their canyons—all features that geologists look for when estimating the age of landforms. The great alluvial fans, which obviously postdate much of the faulting, are nevertheless marked by fault scarps that reflect movement within the last few thousand years.

Along the steep mountain fronts are many examples of large-scale landslides, slides that remain so nearly intact internally that many geologists consider them fault blocks. Best called detachments, they separated from the mountain mass, probably when it was still far below the surface, and slid downhill. They left behind them as they slid many steeply inclined, smoothed and polished slide surfaces, several of which can be seen from the valley floor. Most of the slide surfaces correspond with a weak shale layer at the top of Precambrian rocks. One such surface is at Copper Canyon in the Black Mountains south of Dantes View. Another, near Titus Canyon, shows where clearly layered Paleozoic sedimentary rocks slid off the long dome of the Grapevine Mountains, crumpling, folding, even overturning in the ornate patterns now seen at the edge of the valley.

In the northern part of the national monument, near Emigrant Pass and along the east side of the

Mountains near Racetrack Playa display glistening surfaces where rock slid over rock.

Cottonwood Mountains, other detachment blocks obviously moved *after* mountain uplift. Traveling on cushions of compressed air, they slid rapidly out over alluvial fans and came to rest atop the fans, where their light yellowish color now marks them clearly. In some cases the growing fans have partly engulfed the detached masses.

Death Valley's glaring floor is one of the flattest places on Earth. Yet long-continued tilting, with the east side of the valley sinking more rapidly than the west side, places the lowest spot at Badwater, beneath the sharp face of the Black Mountains. Although in the brilliant desert light it seems to be covered with snow-white salt, much of the playa surface is a mixture of salt, sand, fine silt, and clay, all washed in or blown in from surrounding mountains. During rains, or at times when streams reach as far as the playa, its floor for a time becomes filmed with water; at other times it is hard and crusty. Below the crust are successive layers of fine silt and clay, more salt, and deeper layers of dark plant material. The plant matter tells us that this valley was not always as desolate and barren as it is now, that at times its lakes and marshes supported lush plant growth and no doubt animal life as well.

The surface of the salt pan, smooth when seen from a distance, shows, on closer inspection, many interesting features. It builds from both above and below: from above as flooding streams, containing dissolved salts, spread out over the valley floor and evaporate; from below

as briny groundwater is drawn to the surface by capillary action, there to evaporate. In places the salt pan is marked with shrinkage cracks that form as it dries. Where brine rises through the shrinkage cracks and evaporates, growing salt crystals may tilt the polygons between the cracks. Then, rain and wind reshape the tilted polygons into a jagged surface that all but obscures their origin.

GEOLOGIC HISTORY

Precambrian and Paleozoic Eras. Clues to Death Valley's Precambrian and Paleozoic history come from the bedrock of surrounding ranges. Precambrian rocks include gneiss, schist, granite, and volcanic rocks dated at about 1.8 billion years old, as exposed in the Black Mountains, as well as sedimentary rocks about 1.2 billion years old, which occur in the Funeral Mountains and the Panamint Range. The long erosion at the end of Precambrian time seems not to have attacked this area: The sequence of layered rocks in the Panamint Range seems almost continuous from Precambrian into Cambrian time. Paleozoic sedimentary layers—fossil-bearing limestones, siltstones, and sandstones—were deposited on a broad continental shelf that sloped very gently westward.

Mesozoic Era. In Mesozoic time the seas withdrew as the then west coast of the North American Plate began to rise, uplifted it is thought by collision with microcontinents on the East Pacific

The jagged salt pan surface at Badwater results from evaporation of brine which has seeped to the surface and evaporated.

Plate. As the North American Plate overrode these microcontinents or added them to its own western margin, the vast granite batholiths of the Sierra Nevada pushed upward. Late in the era and far to the northeast the Rocky Mountains began to rise. Volcanism was rife, and small intrusions formed within the Death Valley area.

Cenozoic Era. Mountains in this region were slowly worn down. In Cenozoic time, lakes developed in low spots in the terrain. By their shores lived camels, tapirs, deer, rhinoceroses, titanotheres, several kinds of rodents, and three-toed horses. By perhaps 30 million years ago, the Mesozoic highlands had become low, rolling plains.

Between 13 and 7 million years ago Nevada arched upward, stretching the crust. Violent volcanic eruptions in the Greenwater and Black Mountains, as well as in the southern Panamint Range, blanketed the area with volcanic ash. Earth movements stirred the Death Valley area—domelike uplift, regional arching, movement on strike-slip faults, and sliding of some of the detachments of Paleozoic rocks. Though these movements established mountain ranges separated by intermountain valleys, several million more years were to pass before creation of Death Valley itself.

Around 5 million years ago, volcanic vents within the bounds of the present monument spread more ash across the region. Borax from these volcanic materials would later be washed into lakes that occupied the central part of the valley.

Less than 4 million years ago Death Valley itelf began to open up, its floor gradually dropping as the crust pulled apart. Volcanic ash layers and lava flows along its edges were tilted. Earlier-formed lake deposits were left high and dry on mountain flanks. Along many faults there was so much movement that large blocks of the mountainsides swept downslope and onto the surfaces of new, rapidly developing alluvial fans. Horizontal movement along some faults further widened Death Valley and gave it its overall zigzag shape.

As faulting continued, Death Valley deepened. Erosion tore at bordering ranges. Heavily laden streams built immense alluvial fans along the mountain fronts and conveyed fine silt, sand, and dissolved salts down to the valley floor.

During Pleistocene glacial cycles, when runoff from surrounding areas was greater than it is now, a succession of lakes spread over the valley. At those times vegetation was lush and animal life abundant. In interim warm, dry periods the valley became, as now, a salt-surfaced desert.

About 10,000 years ago the last of the Pleistocene lakes, which geologists have called Lake Manly, dried up. Even more recently, eruptive blasts shaped Ubehebe Crater and its neighbors in the northern part of Death Valley.

The valley continues to subside today. Because sinking is most rapid along its eastern side, the valley floor tilts in that direction. And for the same reason, east-side alluvial fans are not as well developed as those on the west side. Subsiding even as they form, they are soon partly covered with playa deposits.

BEHIND THE SCENES

Artists Drive. The many-colored layers of rhyolite tuff and other volcanic rocks seen along this drive came from explosive eruptions that took place in Greenwater Valley, just east of Death Valley, about 5 million years ago. The reds and pinks, yellows and golds were added later as hot groundwater leached and altered iron minerals in the soft volcanic material. Pastel greens and grays developed as alkaline lake waters reacted chemically with volcanic ash. Dark brown mudstone, which contains less ash and more mud, accumulated in and near lakes along the Furnace Creek Fault. Coarse gravels resembling those of modern alluvial fans show that steep mountains already rose above the lake-filled valley of Pliocene time.

Capped in places with darker basalt, the chaotically colored rocks of Artists Drive slid into their present positions less than 3 million years ago, as detachment blocks that slid downward as the Black Mountains rose and the valley subsided.

Badwater. The lowest point in the Western Hemisphere, Badwater is 86 meters (282 feet) below sea level. Sea level is shown by a white marker 86 meters up on the side of the Black Mountains. A long, narrow fault block composed of highly altered, deformed metamorphic rocks, the Black Mountains rise extremely steeply; their base marks the approximate position of the faults that edge the east side of Badwater Basin.

The deepest, youngest part of Death Valley, this basin first opened up about 3 million years ago. Faulting is still going on, the result of the pulling apart of the Black Mountains to the east and the Panamint Range to the west. Fault scarps only a few thousand years old cross nearby alluvial fans, and shattered rock debris still clings to the cliffs. West of Badwater the valley fill—layer upon layer of salt and mud—is about 2400 meters (8000 feet) thick.

The great Death Valley salt pan, covering more than 500 square kilometers (200 square miles), stretches south and north from Badwater. Salt pools at Badwater show us that the water table is just below the salt pan surface. The water is far saltier than ocean water; on hot, dry summer days salt crystals form on its surface.

With uplift comes erosion. Partly cemented gravel of an old alluvial fan, now raised above the general level, is gradually worn away by water and wind.

Badwater, the lowest point in the Western Hemisphere, lies at the foot of the Black Mountains almost directly below Dantes View. The steep mountain front results from many movements along a still-active fault.

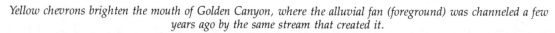

Yellow chevrons brighten the mouth of Golden Canyon, where the alluvial fan (foreground) was channeled a few years ago by the same stream that created it.

Dantes View. The road to Dantes View ascends Furnace Creek Wash to Greenwater Valley, site of explosive Pliocene volcanic eruptions, and then climbs the east side of the Black Mountains. Borax deposits now being mined near Ryan formed as hot groundwater leached borax minerals from Pliocene volcanic rocks on the mesa above the mines.

At Dantes View the surface rock is volcanic. In the cliffs below are Precambrian schist and gneiss dated at about 1.8 billion years old. From this viewpoint Death Valley's dazzling salt pan can be seen in its entirety. Winding across it is the corkscrew course of the Amargosa River, a river that comes to life only when heavy rain falls on its Nevada drainage area. Far to the north, usually out of sight in the desert haze, Salt Creek also feeds into the white expanse. Badwater, the lowest point on the salt pan, is 1750 meters (5750 feet) below, out of sight because of the bulging shape of the mountain face.

Alluvial fans along the base of the Panamint Range, seen across Death Valley, are noticeably larger than those on the east side of the valley. Some surfaces on the great west-side fans are dark with desert varnish developed over tens of thousands of years. Recently active flood channels are lighter in color, not having had time to develop the varnish.

Layered sedimentary rocks stripe the east face of the Panamints. These are Paleozoic rocks that slid *en masse*, as detachments, from the upper parts of that range. Similar sedimentary layers once covered the Black Mountains; they were removed by erosion and by similar detachment sliding, in either case ending up as part of the thick valley fill below the present salt pan.

Devils Golfcourse. Northwest of Badwater is a bed of solid rock salt that is the last vestige of Pleistocene Lake Manly, which dried up about 8000 years ago. This rock salt layer, of almost the same composition as ordinary table salt, is 1 to 2 meters (3 to 6 feet) thick. Now above the water table and out of reach of possible flooding, the salt bed is etched and gullied by rain and wind into a field of jagged pinnacles and spires.

Furnace Creek. Generous streams of fresh, clear water are not what first-time visitors expect to find in Death Valley. Nevertheless, they are here, providing several thousand gallons per minute, enough to nourish the date groves and visitor facilities at Furnace Creek.

The water in these springs comes from inter-mountain basins east of Death Valley. Since groundwater levels there are higher, and since the intervening mountains are honeycombed with fractures and crevices, the water slowly seeps through the mountains and into Death Valley, where it wells up along Grand View Fault. On its trip through the ranges it is warmed to about 40° C (104° F).

Furnace Creek Fault Zone is a strike-slip fault, with horizontal movement amounting to about 30 kilometers (20 miles). In Pliocene time a line of lakes lay along the fault zone. Volcanic ash erupted at that time washed into and filled the lakes in the Furnace Creek area; the water-laid ash now brightens Zabriskie Point and Golden Canyon. It was later covered with brown mudstone and fanglomerate (the conglomerate of alluvial fans), now exposed just east of Furnace Creek Inn and figured on the cover of this book. Fossil leaves and footprints of Pliocene mammals have been found in these rocks.

Golden Canyon and Zabriskie Point. The yellow rocks so well exposed in Golden Canyon and Zabriskie Point are some of the volcanic ash ejected in Pliocene time by volcanoes in the Greenwater Valley area. Washed into Furnace Creek Lake along the Furnace Creek Fault, they were later leached and altered by hot groundwater, in a scenario not unlike that in Yellowstone's hot-spring and geyser basins today. The layers of altered ash, or tuff, are capped by basalt from later eruptions. The whole sequence of rocks has since been tilted and broken by movement along the Furnace Creek Fault, then weathered and eroded into badlands.

Badland topography is not uncommon in desert areas. Necessary ingredients seem to be fine sediment—volcanic ash is ideal—and an arid climate characterized by occasional severe rainstorms—a "cloudburst climate." In such areas there are few plants to hold surface soil, and runoff from sudden storms easily erodes the soft, clayey rock.

Harmony Borax Works. Death Valley's borax results from the same process that created the colorful rocks of Golden Canyon and Zabriskie Point: leaching of volcanic ash by hot groundwater. A chemical salt of the element boron, borax accumulated in lake sediments in layers up to 60 meters (200 feet) thick. Later, after faulting and tilting of these lakebeds, some of the borax was redissolved and transported by streams to the floor of Death Valley. Concentrated by evaporation, it now remains as part of the crust of the Death Valley salt pan.

Borax was scraped from the salt flats of Death Valley, purified in large boilers, and shipped out in huge wagons pulled by teams of 20 mules.

At Harmony Borax Works, borax was scraped from the salt pan surface, and then was redissolved, purified, and dried before being shipped out of Death Valley in wagons hauled by the famous 20-mule teams. Borax is used for ceramic glazes and fertilizer, as well as in the manufacture of glass, fiberglas, solder, water softeners, and pharmaceuticals.

Mosaic Canyon. A walk into Mosaic Canyon is a walk along a detachment fault, where a detached block of Paleozoic sedimentary rocks skidded down the side of Tucki Mountain. The slide surface is a weak zone between Precambrian and Paleozoic rocks. Mosaic Canyon derives its name from well cemented, stream-polished, pebbly conglomerate found at several places in the narrow, flood-carved gorge.

Racetrack Playa. The road to Racetrack Valley is one of the most fascinating in the national monument. It is bordered by small alluvial fans so steep that slumping has made their surfaces

hummocky. Other fans are dissected by streams and reveal their gravelly, cobbly make-up. In places, detachments of scrunched and twisted Paleozoic sedimentary rocks edge their way down polished glide planes. Sand-floored washes and surfaces hard-crusted with desert pavement illustrate desert erosive forms. Prospect pits and small abandoned mines pock some hillsides.

Racetrack Valley's main attraction, however, is found on smooth, silt-floored Racetrack Playa: rocks that slide with the wind.

In this unusual (but not unique) little valley, three special conditions contribute to make the sliding—illogical though it sounds—possible: severe storms, sudden drops in temperature, and strong winds. If occasionally (once in ten years, say) a severe winter storm covered Racetrack Playa with 30 centimeters (a foot) or so of water, and temperatures soon after dropped below freezing (as they often do in storms in this area), ice would form on the water surface and around whatever rocks lay near its margins or out on the

Mosaic Canyon's fluted walls were shaped and polished by stream erosion. The rock is Paleozoic limestone of a detachment block. A cloudburst-created flood left behind the fine, flat-pebbled gravel of the canyon floor.

submerged playa surface. Later, as the ice broke up, some of the rocks might be ice-rafted on floating slabs. Strong winds, such as those that commonly ride the coattails of winter storms, might then push the ice rafts around. Any pieces of rock that projected downward from their ice collars might then drag on the bottom, making long tracklike grooves in the playa mud.

Out on the playa, usually dry and mudcracked, you can see for yourself the tracks left by the moving rocks, and the accumulated bow-wave of mud across their forward edges. Moving rocks are most abundant at the southeast corner of the playa. But no one has ever seen them move.

Sand Dunes. Dunes occur in several parts of Death Valley. The dune field north of Tucki Mountain, which like a natural dam separates northern and southern Death Valley, is the most often visited.

The glaring surfaces of playas are white with evaporite minerals, salt, saltpeter, and gypsum deposited as salt-enriched water evaporated.

Rocks that blow with the wind? Hard to believe, until you see the evidence on Racetrack Playa. Shrinkage cracks formed as the playa dried.

Sand dunes creep downwind as individual grains are bounced up the windward slope and dropped on the leeward.

Much of the dune sand, derived from granite and quartzite exposed in the Cottonwood Mountains, enters Death Valley via Cottonwood Canyon and the large alluvial fan at the mouth of that canyon. Most of the sand is quartz, hard and durable, though there are also grains of feldspar, calcite, and mica. Winnowed from alluvial fan gravels, the sand is swept and bounced along the fan surface or, during particularly strong storms, lifted and carried slightly above the ground. At the same time it is sorted and concentrated: Coarser, heavier sand and pebbles are left behind in windward parts of the dunes, and finer clay grains are blown skyward as dust. The sand accumulates where winds lessen in velocity or, meeting other winds that sweep in from south and north, break up into eddies.

To contribute to sand dune growth, winds must move at velocities greater than 7 or 8 meters per second (15 miles per hour). On the dunes themselves winds of these velocities bounce grains of sand up the windward slopes and drop them at the crests of the dunes, where wind velocity decreases. As crests become oversteepened, sand avalanches down the leeward slopes.

Since sand is swept from windward surfaces and redeposited on leeward faces, you might expect the whole dune field to migrate gradually downwind. However, seasonal variations in wind direction—northerly in winter, southerly in summer—balance out, and these dunes remain in just about the same place from year to year. Individual dune ridges do, however, move and change with time.

"Haystacks" of plant roots were initiated when plants gradually overwhelmed by sand developed longer and longer root systems. Later, the sand was blown away, leaving the free-standing plants high on mounds held in place by their own roots.

The walls of Ubehebe Crater display the fanglomerate (conglomerate of an alluvial fan) that underlies the crater. Cinders blanket this and surrounding areas but do not make up the walls of the steam explosion craters.

Little Ubehebe Crater is the site of another steam explosion.

Look closely at some of the dune surfaces. Walk out among them to see the long windward slopes and the steep leeward faces with their sand avalanches. Filter sand between your fingers, noticing the fineness, the evenness, the roundness of the grains. Notice that large grains accumulate in the ripples. Tracks and trails of animals—kangaroo rats, coyotes, lizards, mice, and beetles—mark dune surfaces. Grasses blowing in the wind leave tidy swirls. Where dune ridges have shifted, plant roots hold together columns of sand, as at the Devils Cornfield.

Titus Canyon. The southern end of the Grapevine Mountains is known for its tremendous and well-displayed detachment block, a mass of Paleozoic (particularly Cambrian and Ordovician) sedimentary rocks which slid down the long dome-shaped core of the Funeral Range. Crumpled, folded, curled over on themselves, some of them shattered and recemented with white calcite, these rocks are visible in the walls of Titus Canyon and on the west face of the Grapevine Mountains. Slickensides, rock surfaces polished by fault movement, in places mark the actual slide surfaces; some can be seen near the sand-floored, rock-walled wash that doubles as a road.

Like other canyons in the ranges surrounding Death Valley, Titus Canyon is steep and narrow—a young canyon, cutting downward rapidly, keeping pace with Death Valley's subsiding floor.

Ubehebe Crater. Among Death Valley's youngest features is a cluster of craters at the north end of Tin Mountain. These are steam-explosion craters, created when hot magma came in contact with groundwater. Here, twelve separate explosions perforated an alluvial fan and spread volcanic ash, cinders, and rock debris over adjacent terrain. Ubehebe Crater, 3000 years old, is the youngest and largest of the cluster. Although in outward appearance these craters resemble cinder cones, the alluvial fan gravels in their walls show that they were cut into rather than superimposed on surrounding terrain.

The smoothly rolling, cinder-covered surface surrounding Ubehebe Crater results from rapidly moving surges of volcanic cinders blasted horizontally by the steam explosions.

Zabriskie Point. See Golden Canyon.

OTHER READING

Chesterman, C. W., 1971. "Volcanism in California," in *California Geology*. California Division of Mines and Geology, vol. 24, no. 8.

Hildreth, Wes, 1976. *Death Valley Geology, Rocks and Faults, Fans and Salts*. Death Valley Natural History Association.

Hunt, C. B., 1975. *Death Valley, Geology, Ecology, Archaeology*. University of California Press.

Kirk, Ruth, 1965 (2nd edition). *Exploring Death Valley*. Stanford University Press.

Troxel, B. W. (editor), 1974. *Guidebook, Death Valley Region*. Geological Society of America, Cordilleran Section.

Gila Cliff Dwellings National Monument

Established: 1907
Size: 2 square kilometers (about 0.8 square miles)
Elevation: 1707 meters (5600 feet) at visitor center
Address: Rt. 11, Box 100, Silver City, New Mexico
  88061

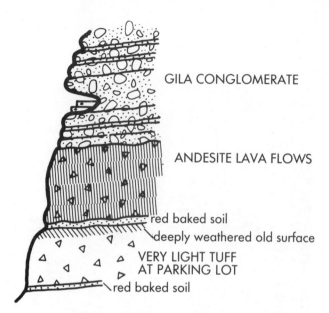

*A stratigraphic section of rocks near the cliff dwellings
shows that caves are cut back into some
soft, poorly consolidated layers of the
Gila Conglomerate.*

STAR FEATURES

• Well off the beaten track, thirteenth century cliff dwellings in natural caves hollowed out of coarse sedimentary rock derived from volcanic material.

• An interpretive program that includes a visitor center and an introductory film.

SETTING THE STAGE

The corrugated upper canyon walls of Cliff Dweller Canyon are composed almost entirely of coarse conglomerate, its pebbles and cobbles, sand and mud reworked from volcanic lava and ash. The conglomerate is part of the Gila Conglomerate, a rock that occurs through much of southwest New Mexico and southeast Arizona (see Tonto National Monument). The same formation is eroded into corrugated cliffs and pinnacles visible from the entrance road, where it overlies the volcanic rocks from which it is derived.

The region as a whole—now the Gila Wilderness—is a large, fairly rugged basin encircling the headwaters of the three forks of the Gila River. In and around the basin is a thick sequence of volcanic rocks, mostly light-colored lava flows, volcanic breccia composed of broken chunks of volcanic rock imbedded in volcanic ash, and thick blankets of light pink tuff.

The Gila Conglomerate and underlying volcanic rocks are sliced by a number of northwest-trending faults. Some fault slivers are raised into horsts, some are dropped down as grabens, in a continuation of the Basin and Range structure of southwest New Mexico and adjacent parts of Arizona.

GEOLOGIC HISTORY

Cenozoic Era. Although we are sure that this area underwent many changes in Precambrian, Paleozoic, and Mesozoic times, rising at times above the sea, or sinking below, sometimes involved in mountain-building, and sometimes resting undisturbed for many millions of years, there is no evidence of this remote history in the national monument area. Here the visible part of the geologic story begins about 30 million years ago, when this part of the earth's crust felt great tension as crustal plates shifted and moved to the west. Faulting occurred, and lava welled up through fault fissures. Occasional explosive volcanic outbursts hurled molten blobs of lava and fine volcanic ash into the air. Falling to earth, the ash became tuff, interlayered in many places with lava flows. Gradually, volcanic cones formed around the most active centers. Successive layers of ash and lava blanketed the region with a mantle ultimately 300 to 400 meters (1000 to 1200 feet) thick.

Corrugated cliffs, shaped by water, frost, and wind, are streaked and stained with lichens.

For 20 to 25 million years, sporadic volcanic activity continued. Between active periods, soil formed and vegetation flourished, only to be destroyed, and the soil baked and reddened, as new lava spread and new ash fell.

After all this volcanism, magma supplies in underground magma chambers were to some extent depleted. The unsupported roofs of the magma chambers, burdened with the weight of thick volcanic layers above them, collapsed to form a broad, irregular caldera. Though its shape has been obscured by later erosion, ring faults show that the caldera covered the area now drained by the three forks of the Gila River. (The viewpoint at mile 29 on New Mexico Highway 15 overlooks this vast mountain-rimmed depression. In last,

dying volcanic gasps, viscous lava surged up in the center of the caldera to form the lava dome on which this viewpoint is located.)

But the volcanic area was not yet completely dead. As erosion took over, washing volcanic debris from highland to lowland, new eruptions occurred, and new clouds of volcanic ash spread and settled over the land. Continued Basin and Range faulting, with a northwest-southeast trend in this area, controlled in a general way the pattern of the landscape: Long slivers rose to form mountains, or sank to form today's intermountain valleys. Some fault blocks were lifted so high that the Gila River and its tributaries, paring them down, laid bare the hard intrusive igneous rock that had cooled slowly within the former magma chambers. (Some of this intrusive rock can be seen from the same viewpoint by looking northwest to Granite Peak, which rises above the canyon of the Gila.)

Hot rocks below the surface still give up some of their heat to hot springs. Hot groundwater has in places leached the rocks; it is responsible for scattered patches of soft white and yellow "rotten" rock similar in origin to the leached rock that gives Yellowstone National Park its name. Watch for some of these patches along the entrance road. Where not altered by groundwater, many of the lava flows contain vesicles, former gas-bubble holes. And in places the vesicles are filled in with moonstone, a silky feldspar mineral valued for its pearly translucence.

Cliff Dweller Canyon, a tributary of the West Fork of the Gila River, is a relatively new feature. It formed where a small stream cut into the flank of an uplifted fault block. Exposure of the Gila Conglomerate initiated formation of shallow alcoves and, later, sizeable caves in weak zones in the rock. Water seeping through porous rock layers gradually removed some of the calcite that bonds the conglomerate. The caves deepened and grew larger as wind and water swept away loose grains of sand, and as blocks and slabs of rock fell from ceilings and walls.

Prehistoric peoples, though they arrived in the area 7000 years ago, did not begin to build in the caves until about 1280 A.D. Although they did not further shape the caves, they freely used fallen and pried-off blocks to build their homes.

OTHER READING

Christiansen, P. W., and Kottlowski, F. E. (editors), 1972. *Mosaic of New Mexico Scenery, Rocks, and History*. Scenic Trips to the Geologic Past, no. 8. New Mexico Bureau of Mines and Mineral Resources.

McFarland, E., 1967. *Forever Frontier: The Gila Cliff Dwellings*. University of New Mexico.

New Mexico Geological Society, 1965. *Guidebook: Southwestern New Mexico*. New Mexico Bureau of Mines and Mineral Resources.

Guadalupe Mountains National Park

Established: 1972
Size: 309 square kilometers (119 square miles)
Elevation: 760 to 2666 meters (2500 to 8749 feet)
Address: 3225 National Parks Highway, Carlsbad, New Mexico 88220

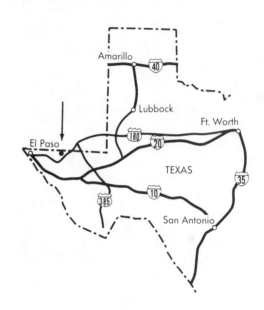

STAR FEATURES

• Canyon and cliff exposures of the world's best understood fossil reef, with its associated lagoon, forereef, and basin deposits.
• A scenic V-shaped fault-block range, with a major fault defining its west side and the reef escarpment outlining its eastern edge.
• Trails and trail leaflets, as well as plans, at this writing, for a visitor center and other interpretive facilities.

SETTING THE STAGE

Lying along the east margin of the Basin and Range Province, the Guadalupe Mountains are unusual in that the sedimentary rocks of which they are composed had their origin in and near a massive limestone reef that in Permian time encircled the shimmering waters of an inland sea. The curving ridge of the Guadalupes—the reef itself—dominates the east side of the range. This escarpment stretches north as far as Carlsbad, New Mexico, but there the massive reef limestone no longer appears at the surface. Underground, it harbors the caverns of Carlsbad Caverns National Park.

National Park Service photo.

The long curve of the Guadalupe Mountains outlines part of a Permian reef. The Capitan Limestone of the reef itself, at upper left, lies on and is fronted with sloping layers of debris broken from it. At the right are flat strata of the Delaware Basin. Lagoon deposits, scarcely visible, form a thin cap on the range.

The entire reef, 550 to 650 kilometers (350 to 400 miles) long and 1.5 to 6 kilometers (1 to 4 miles) wide, outlines an ancient depression, one of three interconnected lobes of a shallow inland sea. The part of the reef that forms the Guadalupe Mountains is only the northwestern bit of the loop-shaped reef, which reappears in the Apache and Glass Mountains. Since it shows up as thick limestone layers in oilwell cores, its buried portions are well known, too. Some wells penetrate hundreds of meters of the massive reef limestone, while others only a short distance away encounter just thin limestone and sandstone layers or deposits of salt and gypsum.

Known as the Capitan Reef, this great mass of limestone was formed biologically, largely as the result of the growth of lime-secreting species of algae and sponges. Many varieties of algae, both fragile upright forms and encrusting head-forming species, contributed to the limestone mass. Other fossil invertebrate animals such as bryozoans, sponges, brachiopods, fusulinids, mollusks, and echinoderms are also common in the reef limestone. Corals are quite rare. (Actually many of today's "coral" reefs are also predominantly formed by algae.)

Geologists studying this reef point out that the sudden apparent increase in calcareous algae and other reef forms indicates an environment sud-

denly favorable for reef growth, probably with a warm climate, warm, clear, freely circulating water, and a not too deep, gradually shelving sea floor. At the optimum depth along the shelving bottom, where sunlight penetrated and yet wave action was not too severe, calcareous algae began to grow in abundance. As they grew, constantly reaching for the currents that brought them needed nutrients, they created a broader level of favorable depth for themselves and for other marine organisms, many of which probably lived in nooks and crannies protected by the algal heads. As individual algae colonies grew, they cemented themselves together, often including fragments of other colonies and shells of marine animals, until a solid mass was created, a true biological reef.

Washed by the waters of the ancient sea, the basinward side of the reef was swept by waves and tides. Bits and pieces and many large blocks of limestone broke off and slid seaward to form an underwater talus slope, the forereef, which merged with basin deposits like those visible along U.S. Highway 62/180. Algae, sponges, and other lime-secreting marine life grew across this rubble toward the currents that brought them the necessities of life, so the reef kept growing seaward. In time it extended several kilometers out over its own debris. All along, it maintained a

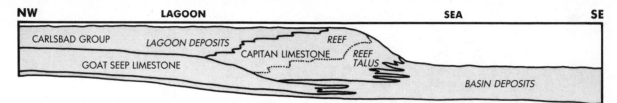

A northwest-southeast cross section shows the Capitan Reef as it was in Permian time. Compare this section with the two below, which show the reef as it is now.

relatively constant width, probably dictated by just how far nutrients were moved by splashing surf, so the lagoon, too, moved seaward. Flat-bedded lagoon limestone in time was deposited above the massive limestone of the reef proper. Thus if you follow a horizontal line from east to west through the mountains, or a vertical one up a canyon wall, you get the same rock sequence: reef talus, reef, and lagoon deposits. Whether portions of the reef were ever out of water, as some parts of most modern reefs are, is not known for sure. Most geologists think the upper surface was continuously submerged, with waves surging among the algal heads, bringing the rich and nourishing chemistry of sea water to the plants and animals living on the reef.

GEOLOGIC HISTORY

Paleozoic and Mesozoic Eras. The most important events in the Guadalupe story happened in Permian time, 290 to 240 million years ago. No older sedimentary rocks are known in the Guadalupe Mountains, but data from oil wells drilled

nearby show that this region had been submerged off and on by shallow Paleozoic seas, in which a succession of layers of marine sandstone, shale, and limestone were deposited.

Early in Permian time, mountains pushed up along what is now the Mexico-United States border. Part of the sea north of the mountains became constricted into a deep basin connected with the open sea by a rather narrow channel. The three-lobed basin is known today as the Permian Basin, a term which has crept from geologic usage into the vernacular of geology-conscious, oil-country Texas. The three-lobed nature of the Permian Basin led to names for its subdivisions, from northeast to southwest the Midland, Delaware, and Marfa Basins. The Midland and Delaware Basins connected with the Marfa Basin and the open sea through a narrow, shallow strait, the Hovey Channel.

It is the central lobe of these three, the Delaware Basin, that was rimmed by the Capitan Reef. Here, during the first half of Permian time, sediments deposited in the Delaware Basin were what one might expect by analogy with modern inland seas: marine limestone near the center of

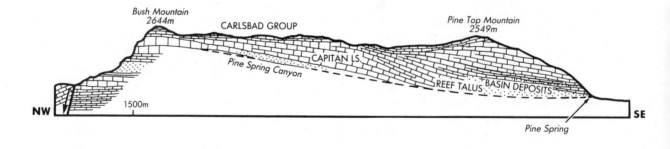

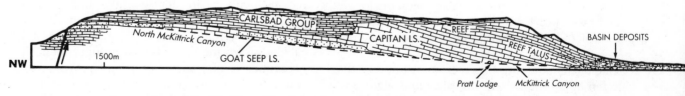

Cross sections of the Guadalupe Mountains show rocks now exposed in the walls of Pine Spring and McKittrick Canyons.

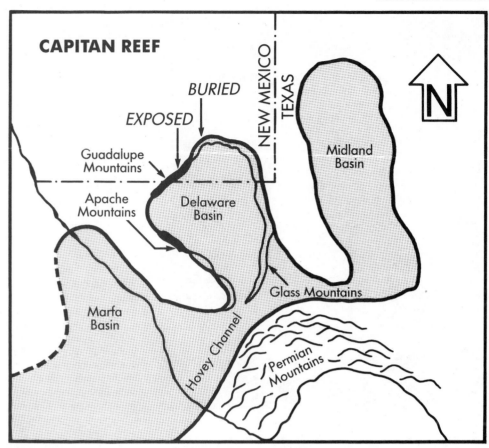

CAPITAN REEF

BURIED

EXPOSED

Guadalupe Mountains

Apache Mountains

Delaware Basin

Marfa Basin

Hovey Channel

Permian Mountains

Glass Mountains

NEW MEXICO / TEXAS

Midland Basin

N

A three-lobed bay connected with the open Permian sea through Hovey Channel. Reef growth finally closed the channel.

the basin, with increasing amounts of sand and silt near shore.

Then, conditions became right for reef growth —whether organic or inorganic—and the reef grew upward rapidly. For millions of years the reef was king, controlling its own growth as well as the kinds of deposits that formed around it. Reef talus accumulated in front of it, lagoon deposits in shallower water behind. Whenever the growing mound partly isolated the lagoon, evaporation led to deposition of evaporite minerals such as anhydrite, gypsum, and in some cases salt.

Finally the growth of the reef, combined probably with localized uplift that increasingly blocked the Hovey Channel, restricted the circulation of water in the Delaware Basin. As the basin's waters began to evaporate faster than they were replenished, the sea became saltier and saltier. Though a shallow channel still connected it with the ocean, highly salty and therefore heavier water was held behind the shallow sill, as saltier water of the Black Sea is held back today. Reef plants and animals died in the high-salt environment, and reef growth came to a halt. Anhydrite and salt ultimately were deposited

clear across the floor of the basin, a vast evaporating pan, burying the reef escarpment and lagoon deposits behind it.

For a brief interval near the end of Permian time, the sea again entered the basin, as it did once more in Cretaceous time. But for all the rest of the Mesozoic and Cenozoic Eras, the land was above the sea.

Cenozoic Era. Late in Tertiary time, about 25 million years ago, earth movements became pronounced. Faulting and uplift began to shape the Basin and Range region we know today. Along the west side of the ancient reef, whose limestones were by then deeply covered with continental sediments, a prominent fault zone developed. The numerous faults of this zone swung northwestward away from the Capitan Reef, so that the present Guadalupe Range is somewhat V-shaped, outlined on the west by faults, on the east by the reef escarpment.

As soon as the Guadalupe fault block rose above the surrounding country, erosion intensified. Mesozoic and Cenozoic sedimentary rocks along the top of the rising range were the first to go. Movement on the fault continued into Quaternary time, and erosion bit ever more deeply.

The rugged prow of El Capitan overlooks flat-lying basin deposits. Guadalupe Peak, higher than El Capitan, is the highest point in the park and in Texas. Joints in the reef limestone provide ready avenues for groundwater solution.

In McKittrick Canyon a small but permanent stream bears little resemblance to the Pleistocene torrents that originally carved the canyons here. The stream flows alternately on and under the surface, and is so heavily charged with calcium carbonate that it deposits travertine.

Many geologic details, including cavernous limestone and a diagonal fault, stand out in the cliffs of Boquillas Canyon.

The granite of Grapevine Hills, cut by many joints, weathers into rounded rock masses.

The South Rim of the Chisos Mountains consists of a thick, resistant lava flow. As a much softer layer of volcanic ash below the flow erodes, the undermined lava breaks away along vertical joints.

Big Bend National Park

Elephant Tusk, the hard igneous rock of a volcano's conduit,
rises above the gravel-covered desert floor.

Soft, easily eroded layers of volcanic ash are exposed in the
Painted Desert south of Maverick.

Zabriskie Point's patterned yellow ridges are composed of volcanic ash deposited in a Pliocene lake and later leached by hot groundwater. A resistant lava flow caps the pointed peak.

Circular pools of clear but salty water mark the surface of the salt pan at Devils Golfcourse.

Freshwater springs are not what newcomers expect of Death Valley. Tule Spring, on the west side of the valley, is fed from the Panamint Range.

Death Valley National Monument

Snow-sprinkled Telescope Peak towers above detachments of Paleozoic rock that form much of the Panamint Range's eastern slope.

Harlequin patchwork along Artists Drive results from detachment faulting.

True to its name, Salt Creek has a salt content few fish can tolerate.

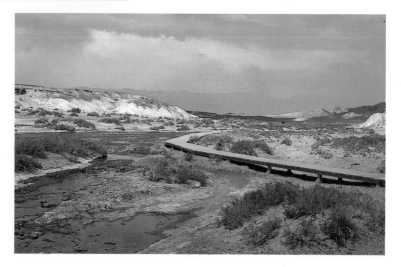

Death Valley National Monument

Over thousands of years, desert varnish darkens an alluvial fan. Light parts of the fan result from relatively recent floods.

Organ Pipe Cactus National Monument

Above and right, *the Ajo Range is banded with Tertiary tuff and lava flows. Basin and Range faulting later lifted the range and tilted its volcanic rocks.*

Organ Pipe Cactus National Monument

Rock arches form as slender blades of rock are attacked on both sides by rain, wind, and freezing of moisture in narrow cracks.

The quartz vein at Golden Bell Mine contained only marginal amounts of gold. Such veins form where fluids in slowly cooling igneous rock seep into cracks and fissures.

Salinas National Monument

Left and below, the rock church at Abo was built of local rock — the Abo Sandstone. The same rock forms resistant ridges across a nearby wash.

Selenite crystals west of Lake Lucero furnish an ongoing supply of gypsum sand.

Sand blown up a windward dune slope eventually avalanches down the leeward face, shown here. Except for occasional wind reversals, the White Sands dunes build northeastward.

A strong southwest wind whirls gypsum sand and dust toward the White Sands' dunes.

White Sands National Monument

Above and right, *interdune areas reveal the ghostly patterns of dunes that have come and gone. As dunes shift, new areas become interdunes.*

National Park Service photo.

On the north wall of McKittrick Canyon, massive reef limestone overlies the sloping rubble of reef talus. Small caves in the reef rock are reputed to have yielded nuggets of gold, a geologic impossibility in rocks of this kind.

Step-like scarps on alluvial fans along the west side of the Guadalupes show us that faulting is still going on, so erosion will continue to wear down the range for some time to come.

As the Pleistocene Epoch brought great ice sheets to much of North America and Europe, cycles of increased rainfall characterized this area. Lakes appeared in many of the adjacent valleys, and increased erosion during the rainy cycles brought about the rugged topography, steep canyons, and many caves and caverns that now characterize the Guadalupes. Because the thick, inflexible masses of reef limestone had cracked, either as the reef and underlying sedimentary layers compacted, or as the mountains were lifted, rainwater and groundwater could penetrate the massive limestone. Many of the present features of this area were brought about by solution rather than mechanical erosion.

BEHIND THE SCENES

El Capitan. Visible from afar, El Capitan's majestic prow, a landmark for early settlers, towers above U.S. Highway 62/180. At the south end of the Guadalupe Mountains, it rises where the reef escarpment on the east side of the range is cut off by the fault scarp of the west side. Thick, light-colored, poorly stratified limestone of its towering cliffs is part of the Capitan Reef. A trail leads to the summit of Guadalupe Peak, just north of El Capitan.

McKittrick Canyon. Offering views of a near-perfect geologic cross section of the Capitan reef, the trail up McKittrick Canyon follows the bouldery stream deposits that characterize most arid-climate canyons. The canyon cuts through the basin limestone and reef talus near the canyon entrance, and soon reaches the base of the massive cliffs that are the reef proper. A little farther upstream the canyon walls reveal horizontally bedded backreef or lagoon deposits. Sloping beds of chaotic, jumbled blocks, recemented with limestone, are visible in the steeply sloping reef talus. Southeastward, these rocks interlayer with the thin, slabby basin deposits that appear along the highway.

In upper reaches of the canyon, beyond Pratt Lodge, thinly bedded lagoon deposits lie above the massive reef rock. Many fossils can be seen in cobbles and boulders of the present streambed.

A guide leaflet to this canyon points out some of the geologic features, including small caves, travertine streambed deposits, fossils, and major features of the Capitan reef.

Pine Top and Bush Mountain Trails. These two trails, together totaling 8.4 kilometers (5.2 miles), provide a good cross section of the reef. Starting at the base of the Capitan reef talus, the Pine Top Trail climbs through it and through the massive limestone of the reef proper, onto an upland surface of lagoon deposits. Examine the reef as you climb through it. Notice the broken and recemented fragments of the reef talus, the massive, virtually unlayered limestone of the reef itself, occasional bands of chert nodules, the tight-packed concentric arcs of ancient algal heads and sponges, as well as joints, small caves, and other erosional features.

From Pine Top Mountain and from many other vantage points along Bush Mountain Trail, the reef escarpment can be seen stretching southwest to Guadalupe Peak and northeast toward Carlsbad. From Bush Mountain, the western faulted side of the range converges southward with the reef escarpment, dropping off sharply to the glimmering white playas of Salt Flats far below.

OTHER READING

Barnett, John, no date. *Guadalupe Mountains National Park, its Story and its Scenery.* Carlsbad Caverns Natural History Association.

Jagnow, D. H., 1979. *Cavern Development in the Guadalupe Mountains.* Cave Research Foundation, Columbus, Ohio.

Kurtz, D., and Goran, W. D., 1978. *Trails of the Guadalupes.* Environmental Associates, Champaign, Ill.

Joshua Tree National Monument

Established: 1936
Size: 2266 square kilometers (875 square miles)
Elevation: 268 to 1772 meters (880 to 5814 feet)
Address: 74485 National Monument Drive, Twentynine Palms, California 92277

STAR FEATURES

• Fault-block ranges whose straight, steep mountain fronts and faceted ridges tell of relatively recent fault movement.

• Rocky wonderlands of jointed, weathered granite, and the older granite core of a special type of range recognized elsewhere in the Southwest.

• A viewpoint looking down into Coachella Valley, a graben created by movement along two branches of the San Andreas Fault. Across the valley, the San Jacinto Mountains consist of a one-time island caught up and added to North America as it drifted west over the Pacific Plate.

• Many examples of the influence of a desert climate on plants, animals, and the rocks themselves. The national monument spans the boundary between the Mojave and Colorado Deserts, and bears characteristics of both.

• Desert oases whose graceful palms are nourished by spring water reaching the surface along faults.

• A self-guiding geologic motor tour, many good trails (some with guide leaflets), visitor centers, guided walks and tours.

SETTING THE STAGE

This national monument straddles the boundary between the Mojave Desert on the north, with an average elevation of about 800 meters (2500 feet) above sea level, and the much lower—near or below sea level—Colorado Desert on the south. Though different in their elevation and hence their biologic features, the two deserts are quite similar geologically. Central valleys of the national monument are surrounded by typical fault-block ranges—the Pinto, Little San Bernardino, Eagle, Hexie, and Coxcomb Mountains. Of these, only the Coxcomb Mountains at the far eastern end of the national monument follow the regional north-south trend; the others curve around to line up with California's east-west Transverse Ranges, of which they form the eastern tip.

As elsewhere in the desert southwest, most of the fault-block ranges rose between 15 and 11 million years ago. Some are still rising, each movement marked with perceived or recorded earthquake activity. Because of their relative youth, mountain fronts project steeply from the valley floors, and display the cut-off or faceted ridges that indicate recent faulting.

Most of these ranges are composed of Mesozoic granite. In places the granite contains unusually large feldspar crystals imbedded in a finer matrix. The rock may also be intricately crisscrossed with white veins, as in the illustration below.

The molten magma which was to become this granite pushed or melted its way upward either in large masses or in narrower, branching sheets that cooled and hardened far below the surface. Later pushed up by Basin and Range faulting, the granite has been carved by erosion into rugged mountain masses and large, picturesque rockpiles.

In the southwest part of the monument, the Little San Bernardino Mountains display a different geologic picture, a pattern seen in a number of desert ranges in southern California and southern Arizona—the type of structure known as metamorphic core complexes (see also Lehman Caves and Saguaro National Monuments). These mountains have huge central domes of granite and/or metamorphic rock that often but not al-

In this intrusive rock, large white feldspar crystals are scattered in a finer-grained background. Two white veins that also mark this rock formed from fluids filtering into shrinkage cracks as the rock slowly cooled.

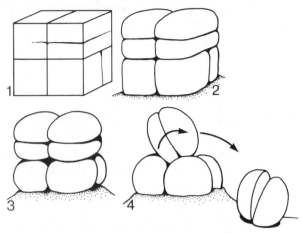

Massive granite broken along parallel joints slowly weathers into piles of rounded residual boulders. Quartz and feldspar grains weathered from the granite accumulate around the bases of the piles.

ways give a recognizable dome shape to the profiles of the individual mountains. The domes are crusted with metamorphic rocks that seem to be squashed and stretched-out versions of the core rock, and that form concentric shells across the domes. Mineral grains in the shells are flattened, sheared, and rotated, so that the metamorphic shell appears streaked or foliated (leaflike). Around some such ranges are thin wedges of Paleozoic and Mesozoic rocks that appear to have moved down the flanks of the rising ranges long before the cores were exposed by erosion.

Geologists believe now that the surfaces of such ranges are great low-angle extension faults formed 30 million years ago as the crust in the Basin and Range area was stretched to the breaking point. Along the fault surfaces, many of which coincide with the tops of large intrusions,

A palm-shaded desert oasis survives on a trickle of water seeping through joints and faults.

Gnarled patterns of Precambrian gneiss show that once-layered sedimentary rock has been subjected to immense pressures and to temperatures high enough to cause recrystallization.

The Pinto Mountains meet the floor of Pinto Valley abruptly, along the line of a fault. A band of dark shrubbery shows that water is near the surface there. Prospect pits and a few small mines indicate mineral enrichment along the fault as well.

lower-plate rocks were crushed and mineral grains were rotated and in many cases reconstituted, so that the rock took on the gneisslike character we see today. The wedges of Paleozoic and Mesozoic rocks that characterize the foothills of some such ranges are parts of the upper plate, above the fault.

In this monument as elsewhere in the southwest deserts, faults that edge fault-block ranges are concealed beneath broad gravel-and-sand alluvial fans. The dramatic south side of the Pinto Mountains, however, meets the valley right at the line of a fault, as does the south side of the Little San Bernardino Mountains. Pinto Wash, at the base of the Pinto Mountains, is for part of its course far on the north side of Pinto Basin, and with the straight mountain face and its faceted ridges demonstrates beautifully the relationship between basins and ranges and the faults that create them.

The spectacular rockpiles of this monument are composed of Mesozoic granite similar to that in most of the ranges. It's easy to see that this pink-

ish, coarse-grained rock is cut into blocks by several sets of parallel joints. Weathering along the joints, with gradual rounding of once-angular blocks, dictated the shape of the individual rocks. Though often called "boulder piles," these rocks don't really conform to the definition of boulders: They are not completely detached from their neighbors, they have not been transported by running water, and they are not rounded by abrasion in the course of transport—all necessary for true boulders.

The rockpiles are surrounded by coarse, grainy sand containing both quartz and feldspar crystals, products of the disintegration of the granite. A close look at the rocks will show you the many more or less rectangular, whitish feldspar crystals, and the clear, glassy grains of quartz. Dark flecks are hornblende and biotite (black mica). On weathered surfaces, iron in these dark minerals has oxidized, giving the rocks their characteristic tan to pink color.

In a number of places within the park there are sharp, easy-to-see contacts between the light-

Although many joints meet at almost right angles, weathering as it proceeds inward from the sides and corners creates more or less spherical rock masses, in a process known as spheroidal weathering.

colored granite and older, darker gneiss. Perhaps the easiest to get to is at Indian Cove near Twentynine Palms. There one can climb along the actual contact for some distance.

The granite of both ranges and rockpiles is marked with many dikes—some quite thick, others pencil-thin. Most of the dikes are light-colored, and are made up almost exclusively of white feldspar. Dikes of this kind form when intrusions cool and shrink, when not-yet-crystallized fluids from the mushy, slowly cooling mass seep into shrinkage cracks, there to harden eventually into dikes.

It's well worthwhile to take the geologic auto tour (a brochure is available at visitor centers) and also to drive to Keys View to look down on the Coachella Valley and across the valley to the San Jacinto Mountains and San Jacinto Peak, 3508 meters (11,502 feet) high. The Coachella Valley ranges in elevation from 300 meters (1000 feet) at its northwest end to well below sea level southeast of Indio, where it merges with the Salton Sea Graben. The Salton Sea, for which the graben is named, has a water surface 72 meters (235 feet) below sea level. It came into being between 1904 and 1907, when the Colorado River overflowed through manmade openings in its natural levee. For a time the sea became smaller as its water evaporated under the fierce desert sun. But now the sea's level remains about constant, kept in balance by irrigation water introduced into the Imperial Valley.

The Coachella Valley-Salton Sea Graben extends south into the Gulf of California. It is a true rift valley, a particularly long, down-dropped sliver between two deep-reaching, more or less parallel branches of the San Andreas Fault. The continent is splitting apart here. If movement along the faults continues (and there is no sign of its letting up) the Gulf of California will become wider and wider, and will extend farther and farther north, eventually becoming a new ocean basin, as the Atlantic did 65 million years ago. These faults are among California's most active, and movements along them have occurred in historic time, producing ground break-

The residual boulders in Joshua Tree's rockpiles formed by spheroidal weathering of once angular blocks. Here, scaling of surface layers keeps the huge rocks rounded.

At Indian Cove (top) and the west end of the Pinto Mountains (bottom) the contact between dark Precambrian gneiss and light Mesozoic granite shows up clearly. Note the difference in weathering characteristics of these two rock types.

age and fissures, small fault scarps, and, of course, earthquakes.

The Little San Bernardino Mountains are part of a relatively high fault block or horst—not so much lifted as not dropped down like the rift valley—north of the San Andreas fault system and south of Pinto Mountain Fault. The raised block is divided by Pleasant Valley and Pinto Basin, both developed along a graben that runs through the center of the monument.

GEOLOGIC HISTORY

Precambrian Era. Precambrian rocks in the Little San Bernardino Mountains consist of 1.8 billion-year-old gneiss—metamorphic rock that is probably of sedimentary origin, and that was intruded by granite well before the end of Precambrian time. In other parts of this region there are also some younger Precambrian sedimentary rocks, partly altered strata whose sedimentary origins are still quite clear. Long erosion at the end of the era leveled the landscape before Paleozoic seas swept in from the west.

The horizontal groove on the sides of these granite boulders marks a former soil level where soil and surface moisture helped disintegrate the rock.

On Malapai Hill, blocks of basalt-like diabase stream downhill in slow-moving rockslides. Both diabase and basalt tend to form such rockslides, as they break readily along column-forming joints. Intrusion of the diabase occurred some 2 or 3 million years ago.

Paleozoic Era. No Paleozoic rocks have been found in this national monument. Their absence might suggest that the area was high-standing in Paleozoic time, and that sedimentary layers were never deposited across it. However, it seems more likely that thick sequences of both marine and non-marine sedimentary rocks were once here but have been eroded away. Very thick sequences of Paleozoic marine sedimentary rocks do occur in several nearby desert ranges. They show no signs of the near-shore deposits that one would expect to find near an ancient highland.

Mesozoic Era. In Jurassic time the area saw intrusion of the granite that now forms its many rockpiles. Mountains formed by the intrusions underwent a long period of erosion, and were reduced to low, gently rolling hills well before Basin and Range faulting took place.

Cenozoic Era. Erosion continued to have the upper hand for some time after the close of the Mesozoic Era. Then, starting about 25 million years ago, uplift little by little raised the area some 3000 to 4550 meters (10,000 to 15,000 feet). Basin and Range faulting subsequently brought about the sinking of basins between mountain ranges. With faulting came erosion. Debris from the rising mountains, carried down into the subsiding basins, concealed their true depth beneath alluvial fans and valley fill.

The deserts themselves came into existence late in geologic time, when the Transverse Ranges and Sierra Nevada rose high enough to cut off moisture from the west. Late additions to the scenery came in contributions from a number of small volcanic centers, which from time to time emitted lava flows and clouds of fine volcanic ash. Fault movements still continue, particularly along the Pinto Mountain Fault at the north

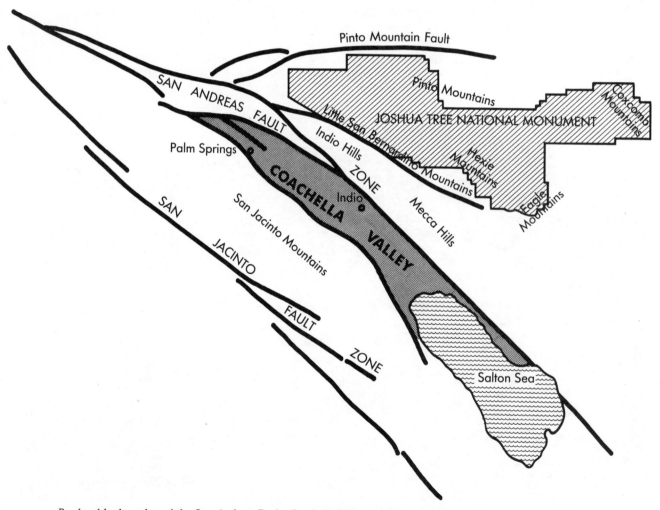

Bordered by branches of the San Andreas Fault, Coachella Valley marks the north end of the Salton Sea Graben. Faults shown here, as well as dozens of smaller ones not shown, have been active in historic time.

A small spring seeps from joints and faults in the granite. This one nourishes Cottonwood Oasis.

boundary of the monument, and along branches of the San Andreas Fault to the south.

OTHER READING

Clark, M. M., 1967. *Map Showing Recently Active Breaks along the San Andreas Fault and Associated Faults between Salton Sea and White-water River-Mission Creek, California.* U.S. Geological Survey Map I–1483.

Larson, Peggy, 1977. *A Sierra Club Naturalist's Guide to the Deserts of the Southwest.* Sierra Club Books, San Francisco.

Trent, D. D., 1977. *The Creation of the Joshua Tree Landscape.* Citrus College, Azusa, California.

Trimble, Stephan, 1979. *Joshua Tree Desert Reflections.* Joshua Tree Natural History Association.

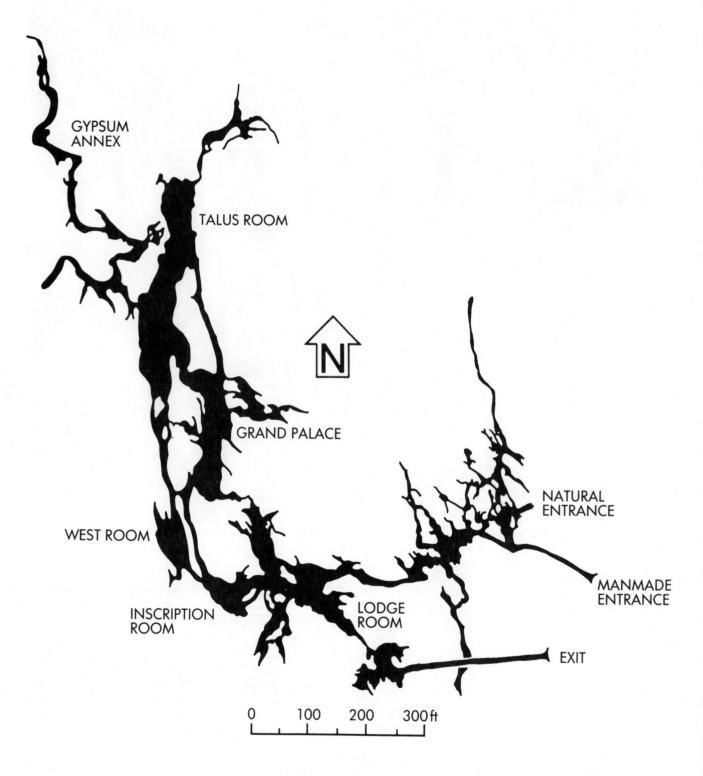

Many of Lehman Cave's rooms and corridors follow northwest-to-southeast joints in the Cambrian limestone.

Lehman Caves National Monument

Established: 1922
Size: 2.6 square kilometers (1 square mile)
Elevation: 2086 meters (6845 feet) at cave entrance
Address: Baker, Nevada 89311

STAR FEATURES

• Underground passages in tilted Cambrian limestone of the Snake Range, prettily decorated with some unusual types of cave ornamentation.

• Visitor center exhibits, cave tours led by Park Service interpreters. In summer, lantern tours explore the main cave and spelunking tours prowl another cavern.

SETTING THE STAGE

High on the slopes of Wheeler Peak in Nevada's Snake Range, Lehman Cave offers a fascinating contrast to nearby snow-capped summits and their surrounding desert basins. The cave, formerly all but invisible at the surface, remains at a cool 10° C (50° F) year-round.

The cavern is a network of underground passageways etched in steeply tilted layers of limestone, rock that saw its beginnings in broad, fairly shallow Cambrian seas more than 500 million years ago. Deposited as limy mud, this rock is soluble in dilute acids such as the carbonic acid present in rainwater and in water that filters down through soil to become groundwater. In solution the calcium carbonate is removed from the new-formed cave channels with the flow of groundwater.

In the solubility of limestone lies the clue to cave formation. Passing through joints and other cracks, groundwater slowly dissolved away the walls of its passageways, widening them little by little over the centuries—a slow process that may have taken hundreds of thousands, even millions, of years.

Joints and faults, important in that they serve as ready-made routes for groundwater flow, control the shape of these and many other caves. The floorplan of Lehman Cave reflects this control: Many of the passageways are parallel, or nearly so, following north-trending and northwest-trending sets of joints and faults. Reflecting the newness of the Basin and Range country, some of the faults are still active, with fault movement averaging up to 5 centimeters (2 inches) per year. The trail through the cave follows the plane of a fault that has shown movement in historic time, offering a rare opportunity to see a fault from the inside!

Solution of limestone takes place most easily where rock is completely saturated. And solution is most rapid just at the water table—the more or less horizontal boundary at the top of the zone of saturation. Because of this condition, the cave is also nearly horizontal, cutting through and across the tilted limestone layers in which it is etched.

The water table is now far lower than it was when the cave was formed, partly because of a general drying of the climate since the Pleistocene Ice Ages and partly because nearby streams have deepened their valleys, which now serve as drainage channels for removal of water.

Lowering of the water table has brought about a second step in cave formation: ornamentation. Rainwater still percolates downward, and as it flows through limestone layers above the cave it picks up calcium carbonate, carrying it, as before, as calcium bicarbonate. When it reaches the cave area, now drained and air-filled, the bicarbonate molecules, no longer under the confining pressure within the rock, release carbon dioxide into the air. The remaining calcium carbonate is deposited in the cave in the form of the flowstone and dripstone ornaments that decorate its rooms and passages.

As in many other caves, the most common ornaments are stalactites, suspended from the cave ceiling like icicles. Tiny rings of calcite, deposited around drops of water dripping from the ceiling, eventually grow into beautiful hollow soda-straw stalactites, some so delicate that light can pass through them. Larger stalactites form as

Courtesy of State of Nevada, Department of Transportation.

Fine soda straw stalactites, knobby stalagmites, and highly ornamented columns developed as calcium carbonate-enriched water dripped from ceiling cracks.

soda straws become plugged with excess calcite, and as water flows down their outside surfaces. Where water drips, splashing, to the floor of the cave, stalagmites form. Where stalactites and stalagmites meet, columns are born, many with second-generation ornaments of their own.

Several less common types of cave ornaments occur in Lehman Cave. Delicate, curving helictites, somewhat of an enigma because they seem to develop with no regard for the downward pull of gravity, are thought to form where tiny pores in the wall or ceiling of the cave admit only miniscule amounts of water—so little that drops do not form. Hydrostatic pressure, rather than gravity, seems to control the formation of helictites. As the pressure forces water through the tiny pores, calcite crystals build up on the surface. The slightly wedged shape of individual crystals causes the growing helictites to curl.

Shield formations are rare in most caves but fairly abundant in Lehman Cave. Each shield consists of two parallel plates, with just a slight crack between them. The plates jut into the cave from narrow fractures in the rock wall or ceiling, and grow as water seeps out of the fractures but is held between the two plates by its own surface tension. Calcium carbonate is deposited at the outer edge, where the water is exposed to the cave air. Here again, hydrostatic pressure rather than gravity controls the flow of water, so the shields can form at almost any angle. If the two sides of a shield separate, water then may drip from their edges, forming ornamental stalactites or draperies.

Calcite lilypads and scalloped pools, cave popcorn, terraces, frozen waterfalls, draperies, and other types of ornamentation are visible in the cave also. The origin of some of these forms is obvious; for more about them see the section on Carlsbad Caverns National Park. Delicate gypsum ornaments occur in certain parts of this cave, but they are so fragile that passageways leading to them are not open to the public.

Some of the ornaments in this cave are tinted with delicate yellows, pinks, and browns. These colors are derived from oxidation of slight amounts of iron occurring in the parent marble or coming from soils and rock layers through which water percolates to reach the cave.

What of the geology of the Snake Range as a whole? It belongs to that class of desert ranges called metamorphic core complexes, ranges that have large, dome-shaped cores of granite or other crystalline rock covered, or partly covered, by shells of metamorphic rock. As with other ranges of this type, there is evidence in this range that the metamorphic characteristics were created by heat and movement along a huge, nearly horizontal extension fault, with movement taking place about 30 million years ago. Precambrian granite below the fault behaved more or less plastically, while predominantly Paleozoic sedimentary rocks above the fault broke along many minor subsidiary faults. Lehman Caves lie in one of the fault slices of Paleozoic rock.

GEOLOGIC HISTORY

Paleozoic Era. In Cambrian time, 570 to 500 million years ago, this area lay just off the west coast of North America, at that time located close to the equator. In the shallow tropical waters that surrounded it, in an enviroment comparable to that around the Bahamas today, layers of limy mud formed from erosion of limy mats secreted by algae. The layers were then gradually covered over with other deposits many thousands of feet thick, and the limy mud was compressed into limestone, the Pole Canyon Limestone of today.

Mesozoic Era. Through the rest of Paleozoic time and into the Mesozoic Era sedimentation continued, with more than 6000 meters (20,000 feet) of sediments accumulating above the Pole Canyon Limestone. By about 70 million years ago, because of the pressure and heat of deep burial, the Pole Canyon Limestone had partially recrystallized, taking on some of the attributes of marble, though because of the fractures and irregularities within it, it never became the true decorative marble of ornamental stone and sculpture.

Toward the end of the Mesozoic Era the area was raised above the sea. The once horizontal rock layers were very gradually bent and thrust over one another, crumpled, broken, and raised into mountains. Fractures and faults created by this movement would eventually determine the pattern of cave formation in the Pole Canyon Limestone.

Cenozoic Era. With the fracturing and uplift came magma—probably not reaching the surface, but rising beneath the layered rocks, its heat causing further changes in the nature of the shale, sandstone, and limestone. Some of the shale at the very bottom of the layered rocks was incorporated into the molten magma. Other shale layers were baked to slaty hardness. Already highly compressed, sandstone was welded into quartzite—the Prospect Mountain Quartzite of Wheeler Peak. The Pole Canyon Limestone may have been further marbleized at this time.

Meantime, erosion attacked the uplifted land, sculpturing the mountains and wearing away their rocks. Then about 15 million years ago another period of mountain-building changed the scene once more, leaving tilted fault-block ranges separated by sunken basins—the Basin and Range topography that we know today. Among the mountains thus formed was Nevada's Snake Range, with the Cambrian quartzite, altered shale, and marbleized limestone, tilted westward by Basin and Range faulting.

Erosion immediately set to work again on these new mountains, carving with running water (and for a time with ice) the present shape of Wheeler Peak and the rest of the range. And erosion worked underground as well, particularly in the Pole Canyon Limestone, as surface water seeped along joints and fault planes to etch out Lehman Cave. The abundant precipitation of Ice Age time, filtering downward along these passageways, raising the water table, no doubt augmented the solution process, so the cave can be thought of as mostly Pleistocene in its basic form.

As the Pleistocene Epoch ended and the climate became drier, less water flowed through the cave. Air-filled, it became the site of deposition rather than solution—deposition in all the shapes and patterns of flowstone and dripstone, of calcite and gypsum crystallization, that one can see there today.

OTHER READING

Halladay, Orlynn J., and Peacock, Var Lynn, 1972. *The Lehman Caves Story.* Lehman Caves Natural History Association.

Montezuma Castle and Tuzigoot National Monuments

Established: Montezuma Castle 1906, Tuzigoot 1939
Size: Montezuma Castle N.M. 3 square kilometers
(1 square mile), Tuzigoot N.M. 23 hectares (58 acres)
Elevation: 975 meters (3200 feet) at Montezuma
Castle Visitor Center
Addresses: Montezuma Castle, P.O. Box 219, Camp
Verde, Arizona 86322. Tuzigoot, P.O. Box 68,
Clarkdale, Arizona 86324

STAR FEATURES

• Prehistoric dwellings in and near natural caves in cliffs of lake-formed limestone.

• A pueblo stronghold on a hilltop of the same limestone.

• An unusual sinkhole caused by collapse of a limestone cavern.

• Nature trails, leaflets, and visitor center displays.

SETTING THE STAGE

The emphasis is archaeological here, but the area's geology is intriguing. The prehistoric Sinagua people who constructed Montezuma Castle, Tuzigoot, and other nearby ruins were the first "geologists" in this area. They identified hard river boulders as suitable for foundations, yet built the upper walls of their pueblos with rough, easily mortared limestone blocks. Near present Camp Verde they mined salt as a trade item as well as for their own use. (Mummified bodies found in a collapsed tunnel reveal that several prehistoric inhabitants died in an early mine disaster.) They also located deposits of soft clay for mortar and pottery, red claystone for jewelry, and hard basalt stream boulders for grinding stones and stone axes. They decorated their bodies and faces with finely ground red hematite (iron oxide), green malachite (an ore of copper), blue azurite (also a copper ore), and white bentonite (a clay derived from volcanic ash). Some of their tools were shaped pieces of diabase from nearby dikes. Others were of obsidian, jasper, and chert. Calcite crystals were used as jewelry.

Montezuma Castle is in a deep natural recess in horizontal layers of chalky limestone, sandstone, mudstone, and conglomerate deposited in the Verde Valley in late Miocene and Pliocene time, 8 to 2 million years ago. At that time, the Verde Valley held a number of small lakes and playas backed up behind fault-created dams at the southeast end of the valley. All told, there are about 600 meters (2000 feet), measured vertically, of lake and playa deposits. Near their eastern end the deposits contain a good deal of volcanic ash. North and northwest of Montezuma Castle and Tuzigoot they are laden with silt brought in by tumbling tributary streams.

The Verde Valley lies between the Mogollon Rim, a steep, dramatically eroded escarpment defining the margin of the Colorado Plateau, and the Black Hills, part of the giant fault-block range that makes up a mountain strip across much of central Arizona.* Bordered along its northern edge by the Verde Fault, the block rose like a huge trap door hinged along its southwest edge. Uplift took place in at least two pulses, each involving many small fault increments accompanied, no doubt, by earthquakes and tremors. Total upward movement eventually exceeded 2000 meters (6000 feet). Most of the sedimentary layers that would correspond with Paleozoic strata in the walls of Grand Canyon, less than 150 kilometers (100 miles) away, are missing from the uplifted block, as are all the Mesozoic strata that once lay above them. Across the valley from Tuzigoot, though, are some Cambrian, Devonian, and Mississippian strata that tell us the Paleozoic seas did indeed cover this area completely.

The Mio-Pliocene lakes, like the present Verde River, received much of their water from calcium carbonate-laden tributaries that drained the limestone surface of the Colorado Plateau. As algae and microscopic animals in the lake water ab-

---

* Recent studies define central Arizona's ranges as merely a more highly lifted part of the Colorado Plateau.

sorbed the calcium carbonate for their shells or skeletons, and later died and sank, limestone accumulated on the lake floor as the Verde Formation. The limestone layers are quite silty and sandy, and usually weather buff rather than white; they form all the irregular grayish and whitish hills in the Verde Valley.

Montezuma Castle overlooks Beaver Creek, a tributary of the Verde River. About 10 kilometers (6 miles) farther upstream is Montezuma Well, a circular sinkhole 143 meters (470 feet) across and about 40 meters (120 feet) deep. Its origin is also related to ancient Lake Verde, for it, too, is surrounded by lake-deposited limestone. Sinkholes are common features in many limestone regions, notably in Kentucky, Puerto Rico, Yugoslavia, Australia, and Mexico (where pre-Columbian priests hurled sacrificial maidens into a similar *cenota*). Groundwater, slightly acid from exposure to the atmosphere and soil, seeps down

High in a cliff of lake limestone, in a cave etched by water and wind, Montezuma Castle housed about 50 people. Others lived below on the floodplain of Beaver Creek.

cracks and crevices in limestone, very slowly dissolving intricate channels and caverns. At first these passageways are water-filled. But if and when the water level drops, unsupported cavern ceilings commonly collapse into sinkholes such as this one.

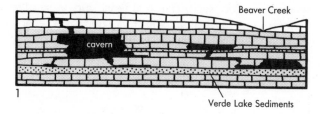

Collapse of an underground cavern caused the formation of Montezuma Well.

High on the slopes of Mingus Mountain, in the Black Hills southwest of Tuzigoot, is the town of Jerome. There, from 1882 to 1954, rich copper-zinc-silver ore was mined—a total of $3.5 billions' worth at today's prices. Jerome's ores were concentrated and smelted in Clarkdale. The now silent mills can be seen from Tuzigoot, across old yellow tailing ponds on the valley floor.

GEOLOGIC HISTORY

Precambrian, Paleozoic, and Mesozoic Eras. The Precambrian, Paleozoic, and Mesozoic history of this area is not revealed in these national monuments, but we can be sure it resembles that of the rest of northern Arizona, and involves:

• Precambrian cycles of sedimentation and volcanism followed by metamorphism, igneous intrusion, and then continent-wide erosion.

• Paleozoic marine and near-shore deposition with interludes of erosion and sand-dune accumulation, but with no volcanism or mountain-building.

• Mesozoic continental deposition followed by one more advance of the sea in Cretaceous time.

Cenozoic Era. Of more immediate concern here is the Cenozoic history of the Verde Valley itself. Late in Cretaceous time and running over into early Cenozoic time, this region felt the impact of

mountain-building that was going on farther north, where the Rocky Mountains were pushing up. Here there was some tilting, some warping of flat-lying Paleozoic sedimentary rocks, and some faulting. For a time, drainage was northward across what is now the Colorado Plateau. Northbound streams deposited gravel on the plateau surface, gravel that contains pebbles of Precambrian rock from highlands in central Arizona.

Then an ancestral Verde River established a southeastward route, slowly carving down through Mesozoic sedimentary rocks between what were to become the Mogollon Rim and the mountains of central Arizona. Fifteen million years ago, in Miocene time, dark, fluid lava flows spread across the region, flows whose remnants now top the Black Hills south of the Verde Valley. Thinner flows were soon to spread over parts of the Colorado Plateau.

As Basin and Range faulting established the desert ranges farther south, the Black Hills and adjacent ranges rose, and minor faulting increased the height of the Mogollon Rim, deepening the partly eroded valley of the Verde River. Coupled with regional uplift over the whole Rocky Mountain-Colorado Plateau region, this new round of uplift sped up erosion, and was responsible in its way for the river power that enabled the Verde to carve down into Paleozoic sedimentary rocks.

Faulting around 8 million years ago also closed up the valley, blocked its outlet, and led to the development of the Verde lakes. Lakes on the valley floor appear to have been shallow, possibly not over 3 meters (10 feet) deep, and at times so salty that salt deposits formed as the water dried up. At other times they were fresh enough to attract the large mammals whose bones and footprints are found in their sediments. Salt crystals show that the climate was arid, but there may have been marshes and greenery around the lakes, tempting animals to linger near the muddy shores. The lakebeds, dated as 8 million to 2 million years old, interlayer in places with datable volcanic rocks, but they also contain fossil bones of mastodons, horses, and turtles, shells of snails and other lake-dwelling shellfish, and tracks of mastodons, camels, cats, and tapirs.

Montezuma Well, Montezuma Castle, and Tuzigoot all owe their sites to slow destruction of the natural dams that had backed up the waters of the Verde River. As a channel was worn down through the rock dams, the lakes gradually drained, leaving behind the chalky, silty lime-

Montezuma Well marks the collapse of an underground cavern in lake limestone of the Verde Formation. Note cliff houses.

stones of the Verde Formation. These still-soft sediments were soon sculptured into hills and valleys by the Verde River and its tributaries. Tributary drainage patterns became established, among them the present course of Beaver Creek. As this small stream cut downward, it chanced to pass fairly close to a large water-filled solution cavern underground in the limestone. Eventually the stream intersected a side passage of this cavern, and partly drained the cavern. No longer water-supported, the thin cavern roof collapsed, and a sinkhole—Montezuma Well—was born. Though blocks of limestone, soil, and plant debris sank into the sinkhole pool, they apparently did not completely block inlet and outlet channels, so water still enters and leaves the sinkhole through submerged openings. There is still underground contact with Beaver Creek, where water from the sink issues in a flowing spring at a rate of 4000 liters (more than 1000 gallons) per minute.

Prehistoric people living near Montezuma Well cut channels to carry this water directly to their fields. In winter the spring water is slightly warmer than normal surface water. It is also richly endowed with calcium carbonate from its passage through twisting limestone caverns. As it cooled, it deposited calcium carbonate on the walls of the ancient irrigation ditches, effectively sealing the stone channels with a waterproof coat that looks like, and is, concrete.

Tuzigoot's isolated hilltop, with its valley-wide vista, was created by erosion, too. Initially the Verde River meandered in a sinuous course across the then nearly flat valley floor, the old lakebed surface. As it slowly cut downward through the lake sediments, its meanders became more and more deeply entrenched. Eventu-

Tuzigoot's high vantage point is a natural hill isolated by a loop of the Verde River.

ally, however, the river cut through one looping meander bend, straightening and shortening its channel and leaving its arcuate older channel as an oxbow lake. You can still see the oxbow meander northwest of Tuzigoot—now a marshy depression largely filled in with stream and pond deposits and plant growth.

Eruption of Sunset Crater on the Colorado Plateau north of here, in 1064 or 1065 A.D., added another page to the prehistory of this region. The eruption of the crater spread moisture-retaining volcanic cinders over nearby parts of the Colorado Plateau. Prehistoric residents, who must have fled in terror during the eruption, soon discovered that the cinders significantly improved crop growth. This discovery brought about a migration of Verde Valley people onto the plateau. Late in the thirteenth century, during a prolonged drought on the Colorado Plateau, migration

seems to have been in the opposite direction, back toward the verdant, irrigated fields of the Verde Valley.

OTHER READING

Jacobs, L. L., and others, 1981. *People of the Verde Valley.* Museum of Northern Arizona, Flagstaff.

Lister, R. H. and Lister, F. C., 1983. *Those who Came Before.* University of Arizona Press, Tucson.

Schroeder, A. H., and Hastings, H. F., 1958. *Montezuma Castle National Monument.* National Park Service Historical Handbook Series no. 27.

As seen from Tuzigoot, the river's early meander bend is now a low-lying curve of marshy ground. Verde Formation cliffs form the background.

Organ Pipe Cactus National Monument

Established: 1937
Size: 1336 square kilometers (516 square miles)
Elevation: 506 meters (1661 feet) at visitor center
Address: Route 1, Box 100, Ajo, Arizona 85321

STAR FEATURES

• Desert fault-block ranges composed of metamorphic, intrusive, and volcanic rocks, with ages ranging from Precambrian to Miocene.

• Desert landforms: sharp, steep mountains, rugged canyons, gently sloping plains, and the dry stream channels known in the Southwest as "washes" and "arroyos."

• A microcontinent picked up as North America drifted southwest over the East Pacific Plate.

• Interesting desert springs frequented by birds and mammals of Mexican affinity.

• Visitor center, illustrated talks, nature walks, desert and mountain trails, and two loop drives with self-guiding pamphlets describing natural features. Back-country roads are usually (but not in all weather) suitable for high-clearance vehicles.

See color section for additional photographs.

SETTING THE STAGE

Where Arizona's desert slopes south toward Mexico, rugged ranges form backdrops for broad intermountain valleys. Along the east side of the monument, visible from the highway or from Ajo Mountain drive, are Miocene volcanic rocks now sliced open by fault movement and erosion. There, dark brown and purple lava and volcanic breccia crudely stripe the Ajo Range, alternating with yellowish layers of volcanic ash. These rocks tell of repeated bursts of violent volcanic activity between 22 million and 14 million years ago, before and during the time when Basin and Range faulting lifted the desert ranges and dropped the valleys between them.

In the central part of the monument is a range nearly as rugged, the Puerto Blanco Mountains. Here are found more volcanic rocks, as well as much older gneiss, schist, granite, and veins of coarse-grained pegmatite. Small prospect pits and a few abandoned mines reflect mineral enrichment, generally associated with dikes or other small intrusions, in this range.

Northwest of the Puerto Blanco Mountains, continuing their northwest-southeast trend, another dark volcanic range stands stark and barren: the Bates Mountains, with Kino Peak, 974 meters (3197 feet), their highest point. In parts of this range fairly young, horizontal basalt flows top the older volcanic rocks, forming flat summit plateaus.

Like other ranges in the Basin and Range region, the Ajo Range, Bates Mountains, and Puerto Blanco Mountains are fault blocks bordered by vertical or steeply inclined normal faults, with sunken valleys between, in the pattern so characteristic of the southwest deserts. Whittled by water and wind, the ranges have shed, through time, great quantities of sand and gravel, partly burying themselves in their own debris. Around each range, gravel-covered piedmonts slope out into the surrounding valleys. Close to the mountains these slopes are erosional, cut into the mass of the mountains themselves, and only thinly covered with gravel. But some distance out from the mountains, beyond the faults that edge the uplifted blocks, are thick, layered deposits of basin fill—gravel, sand, and possibly, hidden from sight, lava, volcanic ash, and lake deposits. The evenness of the piedmont slopes and their gravel veneer make it tricky or impossible to locate visually the exact position of the faults, though the change from thinly covered, beveled rocks of the mountain blocks to thick deposits of basin fill may show up in gullies, and can be located by gravity measurements.

The Ajo Valley, between the Ajo Range and Puerto Blanco Mountains, is typical of the desert's intermountain valleys. From the highway one can see that the gravel-covered piedmonts

Dark volcanic rocks of Twin Peaks rise above the desert floor. Tertiary and Quaternary volcanic rocks are common in the ranges of southwestern Arizona. Many Tertiary flows are tilted because of Basin and Range faulting; most Quaternary flows remain horizontal or nearly so.

Cherioni Wash cuts through Growler Canyon as the only drainage route for the Ajo Valley. Basalt-capped hills flank it on both sides.

slope gently from both ranges toward Cherioni Wash, well over on the west side of the valley. Washes draining both ranges are tributaries to this master wash, which provides the main drainage route through the Bates Mountains. Like many another stream course in this region, this wash and its tributaries are usually bone dry—on the surface at least. After hard mid-winter or summer rains, however, their channels collect water from smaller tributaries and funnel it toward Growler Canyon, where it passes through the Bates Mountains. The stream's course through this canyon seems to be a relic of earlier times, when the mountains were partly covered with lava flows and valley debris. As the mountains continued to rise, and as downstream valleys (such as that of the Sonoyta River south of the Mexican border) were deepened, much of this debris was swept away. The stream, how-ever, more active than at present, kept to its al-ready established course through the Bates Mountains. West of Growler Canyon and the Bates Mountains, Cherioni Wash becomes Growler Wash, and in the desert environment

most or all of its water sinks into the sand of Growler Valley.

Paralleling the north-south faults that edge the Ajo Range and the Puerto Blanco and Bates Mountains, another major north-south fault cuts through the central part of the Bates Mountains, separating Kino Peak from ridges farther to the east. The valley eroded along this fault, easily seen from the northernmost part of the Puerto Blanco loop drive, drains in two directions: north into Growler Wash and southwest toward Cipri-ano Hills. The same fault extends farther south between the Puerto Blanco Mountains and Cipri-ano Hills, possibly all the way to the Mexican border.

Metamorphic rocks and granite are most abundant on the west flank of the Puerto Blanco Mountains, in the Sonoyta Mountains (including Senita Basin) and in the Quitobaquito Hills, all in the southwest quarter of the national monu-ment. Puerto Blanco Drive comes upon these rocks as it curves around the northwest end of the Puerto Blanco Mountains.

Farther south along the loop drive, near Quito-

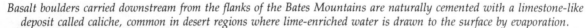

Basalt boulders carried downstream from the flanks of the Bates Mountains are naturally cemented with a limestone-like deposit called caliche, common in desert regions where lime-enriched water is drawn to the surface by evaporation.

baquito Springs, the geology is especially complicated. The Quitobaquito Hills consist of a thick stack of eight sheets of shattered rock that represent a major thrust fault, where movement took place under a heavy load of overlying rocks that have now eroded away.

GEOLOGIC HISTORY

Mesozoic and Cenozoic Eras. Aside from a few patches of Precambrian granite in the far southwestern part of this national monument, rocks exposed here are all Mesozoic or younger. Gneiss and schist found in the Puerto Blanco Mountains are probably Jurassic. They were intruded, in Jurassic and Cretaceous time, by the coarse, light-colored granite that appears along the western part of Puerto Blanco Drive. Granite around Senita Basin in the Sonoyta Mountains is younger, probably Tertiary.

Bonita Well, on Puerto Blanco Drive west of the Puerto Blanco Mountains, taps valley sands and gravels for groundwater. Once used by ranchers, it now attracts animals native to this water-short country.

Most of Organ Pipe Cactus National Monument, along with much of the area between the monument and Tucson, differs in its basic geologic make-up from other parts of the Basin and Range region in that it has seen three major episodes of faulting:

• Push-together thrust faulting in late Cretaceous and early Tertiary time, 70 to 60 million years ago, represented by the many-layered thrust fault in the Quitobaquito Hills.

• Pull-apart faulting about 30 million years ago, with nearly horizontal extension faults, in a pattern seen in many other parts of the Basin and Range region.

• High-angle Basin and Range faulting some 15 to 11 million years ago, responsible for the present ranges.

Superimposed, these faults have made the history of this area unusually complex. A long period of erosion followed each round of faulting. Between the 30-million-year-old extensional faulting and Basin and Range faulting, starting

around 15 million years ago, volcanic eruptions also altered the face of the land. In three long-lasting episodes, three groups of volcanic rocks were produced, each consisting of thick, stubby lava flows alternating with irregular sheets of volcanic ash produced by explosive eruptions. These lava and ash flows, layer upon layer, gradually built up as the thick piles of volcanic rocks that now make up the Ajo Range, the Bates Mountains, and the northern part of the Puerto Blanco Mountains.

The nature of the volcanic outpourings changed over time. At first, the lavas were fairly light in color and fairly thick and sticky. Gradually they became darker and more fluid. The youngest flows were the most fluid, and today form the nearly level summits of the Bates Mountains, as well as the Ajo Range summits of Tillotsen and Diaz peaks.

Basin and Range faulting began about 15 million years ago and lasted for 4 million years. Vertical or nearly vertical faults cut across the two

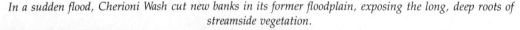

In a sudden flood, Cherioni Wash cut new banks in its former floodplain, exposing the long, deep roots of streamside vegetation.

sets of nearly horizontal faults, as well as across the thick volcanic piles, raising the present ranges and dropping the valleys and basins between them. Erosion of course attacked the mountain blocks as soon as they began to rise; debris from the ranges was carried into the valleys, completely concealing major parts of the old overthrusts and hiding details of the two earlier episodes of faulting.

At first, many of the basins lacked external drainage. When precipitation was adequate, they held lakes. In the Ajo Valley, Cipriano Wash, as it first developed, probably flowed for a time on the upper surface of the youngest lavas, where it found easy passage across what are now the Bates Mountains. Then as the mountains gradually lifted, the stream cut just as gradually downward, incising Growler Canyon, which became the main exit route for all of the Ajo Valley.

Other features of the present landscape evolved at about the same time. The fault that bisects the Bates Mountains, for instance, and that is reponsible for its central valley, developed after the basalt lavas were in place. As gradients steepened, the present rugged canyons and ridges of the Ajo Range succeeded an earlier, less craggy terrain.

The Pleistocene Ice Ages brought changing climates to this region. Rainfall increased, canyons deepened, and alluvial fans widened and merged with each other as immense amounts of rock material were dumped into the intermountain basins. Some basins held freshwater or salty lakes. Mountain slopes may have for a time been thickly forested.

Then, with rising temperatures and decreasing humidity at the end of the Ice Ages, the region slowly took on its present appearance. Lakes dried up; streams became sand-floored washes, flowing only after rains. Plants and animals able to adjust to these changed conditions survived; the others migrated northward or perished.

OTHER READING

Keith, Stanton B., 1971. *Geologic Guidebook 3: Highways of Arizona.* University of Arizona, Tucson.

Saguaro National Monument

Established: 1933
Size: 339 square kilometers (131 square miles)
Elevation at visitor centers: Rincon Unit: 933 meters (3062 feet); Tucson Mountain Unit: 722 meters (2370 feet)
Address: Route 8, Box 695, Tucson, Arizona 85730

STAR FEATURES

• Two fault-block ranges, quite different geologically, one composed of a large mass of intrusive and metamorphic rock, the other largely sedimentary and volcanic. Though their histories tie them together in an unusual way, each range represents a different mountain-building episode.
• Many examples of desert geologic processes.
• Loop drives and hiking trails (some with illustrated leaflets), naturalist-led walks, visitor center displays, and evening programs.

See color pages for additional photographs.

SETTING THE STAGE

Both the Rincon and Tucson Mountain units of this national monument seem at first glance to be confusing medleys of different rock types whose relationships with each other are far from clear. However, both units adjoin or include mountain ranges characteristic of the whole Basin and Range Province. One range—the Rincons—is a metamorphic core complex range composed almost entirely of intrusive and metamorphic rocks. The other—the Tucson Mountains—is

A typical desert wash flows only briefly after a rain. The braided pattern of the stream is characteristic of many sand-floored washes.

made up of much younger sedimentary, intrusive, and volcanic rocks. Let's look at the two units separately:

Rincon Mountain Unit. The Rincon Mountains of the eastern monument unit are far larger than the Tucson Mountains of the western unit. The main mass of the Rincons, its real core, is an immense dome of Precambrian granite shelled with a thick layer of Tertiary gneiss that gives every appearance of having been derived directly from the granite. The dome is three-humped, with summits cresting on Mica Mountain, Tanque Verde Peak, and Rincon Peak, all within the national monument. Prominent mountain spurs, Tanque Verde and Rincon Ridges, extend southwestward from Tanque Verde and Rincon Peaks. Both of these ridges look somewhat like anticlines, but geologists call them antiforms because they don't involve sedimentary rocks and because they aren't caused by folding. The main Rincon dome is duplicated in the Santa Catalina Mountains to the northwest, a range that is continuous geologically with the Rincons, but that

has a high antiform ridge along the south side of its granite dome.

Both the Catalinas and the Rincons are typical metamorphic core complex ranges, similar to a scattering of other such ranges in a band that stretches north to Canada and south into Mexico (see also Joshua Tree and Lehman Caves National Monuments). The Rincons in particular have served as a study site for geologists trying to unravel the strange history of metamorphic core complexes.

The gneiss that surfaces the Rincons is a particularly hard, tough, resistant rock, not very susceptible to erosion. Because of its toughness it still coats most of the Rincon Mountain mass, forming a shell, or carapace, over the granite core of the range. Below the mountain slope, lying against the tough carapace, is a zone of intensely fractured and much more easily eroded gneiss, pale gray—nearly white in places—and composed almost entirely of feldspar crystals dotted with smaller crystals of mica. It is banded with rusty-looking gray schist (the "rust" is iron oxide from decomposing biotite) and in places it

Bordering a tumbling torrent, fractured gneiss reveals streaks and patterns reflecting variations in mineral content.

Rocks highly polished by fault movement catch the morning sun.

is cut by dark green diabase dikes or laced with white quartz veins.

Above both the carapace gneiss and the fractured layer is a broad band of dark, greenish, finely crushed rock topped with a shiny, highly polished surface that is thought to represent a huge extension or pull-apart fault, the Santa Catalina Fault. Horizontal displacement on this fault may have been as great as 50 kilometers (about 30 miles). Around 30 million years ago, extension faults such as this thinned and widened the Earth's crust all up and down the Basin and Range region, adding several hundred kilometers (100 to 200 miles) to its total width.

All the rocks discussed so far lie *below* this great fault and fairly near the main mountain mass. They can be seen along the east part of Cactus Forest Drive. Rocks above or valleyward from the fault (which approximately bisects the loop drive) can be seen along the western side of the loop, in the outer part of the foothills that edge the range. A thin veneer of cobbles, peb-

bles, and coarse sand hides most of the bedrock, but in stream gullies and roadcuts the above-the-fault rocks are exposed. A common rock type is mica schist, a silvery metamorphic rock that outwardly appears to be made up almost entirely of mica grains. A patch of Mississippian, Pennsylvanian, and Permian limestone, also a remnant of the old upper plate, forms the ridge encircled by the south part of the loop drive; another small patch is southwest of the picnic area. (More extensive exposures occur outside the monument along the mountain front. For more on these upper plate rocks, see Tucson Mountain Unit, below.)

Lapping up against the foothills of the Rincons are Tertiary gravel and sand, basin deposits made up of fragments of rocks from the mountain. You can find most of the rock types of the Rincons among the cobbles and pebbles of washes and streambeds here: granite and gneiss, dark fine-grained diabase, silvery cobbles and pebbles of schist, and occasional bits of sed-

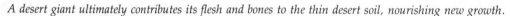

A desert giant ultimately contributes its flesh and bones to the thin desert soil, nourishing new growth.

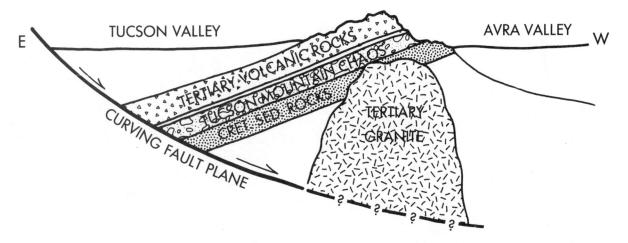

In the Tucson Mountains, volcanic and sedimentary rocks dip eastward. They may represent a tilted block that moved westward along a curving fault.

imentary rock. These rocks occur as well in the desert pavement that veneers the foothills. Two other rock types should be mentioned; both appear in small patches along the north part of the loop road. Andesite and rhyolite, both volcanic rocks, occur as remnants of ash falls and lava flows. Light-colored, strangely shaped feldspar crystals in the andesite give it its geologic name: Turkey Track Andesite.

Tucson Mountain Unit. The west unit of the monument, along the west flank of the Tucson Mountains, is geologically quite different in both rock type and structure. The fault block of the Tucsons tilts eastward, exposing on its western face rocks that range in age and type from billion-year-old Precambrian schist to 20-million-year-old Miocene lava flows. Within the boundaries of the west unit, however, almost all of the rocks are Mesozoic or younger. The lowest and oldest are the bright red sandstone and mudstone layers that emblazon the hills behind the visitor center, collectively called the Recreation Redbeds. They are now thought to be of Triassic age, and related to red Triassic sedimentary rocks elsewhere in Arizona. Above them are layers of a coarse sandstone known as the Amole Arkose. Arkose differs from other sandstone in that it contains abundant feldspar grains. Because feldspar minerals break apart and decompose much more rapidly than quartz grains, we know that arkoses—this unit among

them—are deposited quite near the original sources of their sand.

Above the Amole Arkose is a rock unit known as the Tucson Mountain Chaos, a strange and enigmatic unit that brings together large blocks of Precambrian schist, Paleozoic limestone, Mesozoic sedimentary rocks, and Tertiary volcanic rocks, all bound together by a matrix of sandstone and tuff. Most rock units, formations, are given names that indicate their dominant rock type. In the case of the Tucson Mountain Chaos, it was obviously difficult to decide which rock was dominant and what its origins were. The unit as a whole may be volcanic, with surging lava having picked up large chunks of whatever rocks walled its underground conduits. Or lava and ash may have flowed across a bouldery surface, picking up some boulders, flowing in around others. Parts of the unit may be flood-deposited, with the floods originating during or just after violent volcanic eruptions—like the Toutle River flood after the 1980 Mount St. Helens eruption.

The Cat Mountain Rhyolite, a layer of tuff welded together by its own heat, is superimposed on the Tucson Mountain Chaos. Most exposures of this unit are outside the national monument.

All of these formations are much faulted, and all are intruded by the granite that now forms Amole and Wasson Peaks and appears as well in the western part of the monument near the scenic

Darkened by desert varnish, the Cretaceous volcanic rocks of the Tucson Mountains show many finely spaced parallel fractures.

drive. This granite— a coarse, light tan rock that weathers into rounded knobs and coarse desert sand—may be the intrusive equivalent of the Cat Mountain Rhyolite, which is of nearly the same age and mineral composition.

GEOLOGIC HISTORY

Precambrian Era. What do these two ranges—the Rincons and the Tucsons—tell us about the geologic history of southern Arizona? The story is patchy at best, and involves many different geologic processes. The oldest rocks are in the core of the Rincon Range, where about 1.6 billion years ago, in Precambrian time, huge bodies of magma rose partway through the crust and slowly cooled into granite. Possibly at this time the rocks were also sheared in such a way that their crystals lined up, giving them some of the grain that exists in them today.

Paleozoic Era. From late Precambrian until Cretaceous time, roughly from a billion to 100 million years ago, the region was fairly stable, rising and falling as a unit but not tilted, squeezed, stretched, or broken. Shallow seas that first crept across southern Arizona in Cambrian time paid repeated visits during the rest of the Paleozoic Era, alternating with long above-sea-level erosion intervals.

The few Paleozoic rocks exposed in the northern Tucson Mountains resemble those of the Rincon foothills and of several other southern Arizona ranges. Similar limestones occur in several of the small hills just west of the Tucson Mountains, as well.

Mesozoic Era. During much of the Mesozoic Era this area remained above the sea and therefore experienced more erosion than deposition. Nevertheless, some sedimentary rocks originated during that era: the brilliant sandstones and mudstones of the Recreation Redbeds, which were deposited on a Triassic river floodplain or delta, and the Amole Arkose, originally part of an alluvial fan. Mesozoic volcanic rocks include parts of the Tucson Mountain Chaos. The confused conglomerate of the Tucson Mountain

Chaos shows a change in matrix from sand near the bottom to volcanic ash near the top, reflecting the increase in volcanism that characterizes both the Cretaceous and Tertiary Periods in this area.

Late in Mesozoic time, as North America broke away from Europe and began to drift westward over the East Pacific Plate, the region rose well above the sea. As the East Pacific Plate was forced downward below the continent's edge, some of these rock units were bent and folded. At the same time, magma surged upward to form intrusions of many sizes, including the Amole Peak-Wasson Peak granite, and to escape as lava flows or bursts of volcanic ash such as those which solidified into the Cat Mountain Rhyolite. To some extent the edge of the continent buckled and broke. Bits and pieces of new land, formerly islands or microcontinents, were added to the edge of North America, each addition placing the coastline farther west and southwest.

Cenozoic Era. Caught between the west-drifting North American Plate and the downward-plunging East Pacific Plate, this region continued to bend and buckle. In places, magma rose to form new intrusions or to cause new volcanic outbursts. Then, about 30 million years ago, horizontal compression gave way to tension and a stretching out of the crust in this area. As a result of this tension, the Basin and Range region seems to have fragmented into thin, nearly horizontal slices that slid across each other like a toppled stack of pancakes. In the Rincon-Catalina area, rocks of an upper slice—above a gently inclined, curving fault—moved across great granite intrusions of Precambrian and early Tertiary age, creating the zone of crushed, altered, finely sheared rock that we have seen in and on the Rincons. In the process, partial melting reset the atomic clocks of the altered rock, so that they now read "Tertiary."

Movement along the great fault may have continued for 10 million years; it gradually uncovered large areas of once deeply buried rocks. As the heavy load of rock above them was removed, the massive granites that would become the Rincon and Catalina mountains became buoyant, and domed upward. Shielded from erosion by their tough shells of gneiss, they retained their domed shape, even as softer, highly fractured, less resistant parts of the metamorphic shells were stripped away.

What became of the upper slice, the upper plate above the fault? It broke up, parts of it remaining around the mountain core, parts moving westward or southwestward across the gentle curve of the extension fault. Because of the curve of the fault, its once-horizontal rocks tilted backward toward the area from which they had come, a tilt that can be seen today in the east-dipping rocks of the Tucson Mountains. So the surprise ending of the story of this fault is that the Tucson Mountains may well be the upper plate that once covered the Rincons, and the Tucson Mountain Unit of the national monument may once have been superimposed on the Rincon Mountain Unit!

One more round of faulting completed the picture—the Basin and Range faulting of 15 to 11 million years ago. During this episode, the Tucson Basin sank relative to its bordering ranges. The Tucson Mountains—whether or not they are the old upper slice—became a fault-block range in their own right. Since then, the mountains have been further shaped by erosion, and the Tucson Basin has gradually filled with thousands of meters of debris worn from the mountains—the same picture we see in other mountain-fringed basins of the Southwest.

Desert processes going on today constantly give new touches to the geologic scene. The broad slopes around the mountains, the bare rock surfaces, the pebbly desert pavement, the steep-walled gullies, are all youthful features. During parts of Tertiary and Quaternary time this region was more heavily vegetated than at present. In Pleistocene time temperatures dropped, and during the rainy cycles that accompanied northern glaciation, there was enough rainfall here to augment erosional stripping of the mountains and filling in of the valleys. Lakes and shallow playas may have appeared on the valley floor at that time. Then as the Ice Ages closed and the climate became warmer and drier, lake water evaporated and the basin slowly became the desert you see today—a baby by geologic standards.

OTHER READING

Drewes, Harald, 1977. *Geologic Map and Sections of the Rincon Valley Quadrangle, Pima County, Arizona.* U.S. Geological Survey Map I–997.

Shelton, Napier, 1972. *Saguaro National Monument, Arizona.* U.S. National Park Service.

Salinas National Monument

Established: 1909 as Gran Quivira National Monument; 1980 as Salinas National Monument

Size: 4.2 square kilometers (1.6 square miles) total for three sites

Elevation: 1980 meters (6495 feet) at monument headquarters in Mountainair

Address: P.O. Box 496, Mountainair, New Mexico 87036

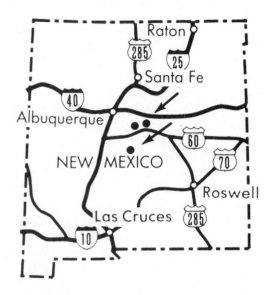

STAR FEATURES

• Ruins of three prehistoric/historic pueblo Indian villages and their five early seventeenth century Spanish mission churches.

• Visitor centers and self-guiding trails through the ruins.

• A coming together of geology, archaeology, and history in the ruins, in nearby evidence of an ancient lake, and in several salt lakes used by both Indians and Spanish settlers as sources for salt.

SETTING THE STAGE

Just east of the Manzano Range, in a broad valley known as the Estancia Basin, are three archaeological sites, Quarai, Abo, and Gran Quivira, now combined into one national monument. The three villages, or *pueblos*, built by pueblo Indians related to those of the Rio Grande Valley, date back to the twelfth, thirteenth, and fourteenth centuries. With the coming of Spanish padres in the early years of the seventeenth century, mission churches and convents were added to these villages.

Both churches and surrounding pueblos, now partly excavated and stabilized, were built of lo-

Directed by Spanish padres, Indians of the Abo village used naturally squared blocks of Abo Sandstone to construct a mission church.

National Park Service photo.

Their surrounding village nearly erased by the hand of time, Gran Quivira's churches and convents rise from the crest of a hill of San Andres Limestone.

cal rock—San Andres Limestone at Gran Quivira, red sandstone from the Abo Formation at Abo and Quarai. The thinness of individual beds in these formations, and parallel joints in sets at nearly right angles to the bedding and to each other, made it fairly easy for the early builders to shape the stone even with primitive tools.

For details of the geology within the national monument, let's look separately at the three sites:

Gran Quivira. The ruins at Gran Quivira date from the twelfth century. (A pithouse village nearby may date back to the eighth century.) The pueblo served as a major trade center between Indians of the Great Plains and the Rio Grande Valley, and later between the Indian and Spanish settlements farther north and west. The ruins are located on a hill of San Andres Limestone, an eastward extension of Chupadera Mesa. Mantled

with a thin layer of sediment, the rock layers on both hill and mesa are almost horizontal. The rock is an impure marine limestone. Here it contains a great deal of gypsum—enough to make both surface and subsurface water of poor quality. Water for the village was obtained from rooftop runoff and from walk-in wells—32 of them, according to Spanish historical documents. Rock and clay cisterns were used for water storage. There are no reliable springs near Gran Quivira. (Visitor center water comes from a 600-foot drilled well.)

The San Andres Limestone seen here is of the same age as the Kaibab Limestone on the rim of Grand Canyon. These limestones represent soft, limy muds probably made up of calcareous shells of microscopic animals and plants, deposited in shallow seas of a continental shelf. The area in which they were deposited was limited on the north by eroded highlands of the Ancestral

Blocks of Permian limestone were used in building Gran Quivira.

A slope-ledge-slope pattern results from differential weathering of soft shale and harder sandstone layers in the Abo Formation. Abo ruin is at lower left.

Rocky Mountains, developed in Pennsylvanian time in approximately the position of today's Rockies.

The terrain drops off north and east of Gran Quivira, with sand hills, remains of old sand dunes, filling the depression between the ruins and the Gallinas Mountains farther east. In Pleistocene time, and probably for short periods when the site was occupied, this natural depression held a succession of small lakes. Later, the fine, loamy sand deposited in the lakes served the Indians well: Corn cultivated on these soils could have supported a sizeable population. In the nineteenth and twentieth centuries, however, deep plowing played havoc with the soils of pueblo times, and the basins now continue to deepen as wind erodes away the fine lake sediments.

Abo. The pueblo of Abo also served as a large population center, with perhaps 2000 inhabitants. Close to the Manzano Mountains, it was fortunate in having a more dependable water supply than Gran Quivira. The ruins occupy a valley between ridges of Yeso Formation to the east, and red rocks of the Abo Formation to the west. The Abo Formation is about 90 meters (300 feet) thick here. Early Permian in age, the unit includes many alternating layers of red shale, fine reddish brown sandstone, and coarse gray or dusky red sandstone and shale. There is much evidence, including lenses of sandstone deposited as channels and bars, and overbank deposits formed during floods, to indicate that the Abo Formation was deposited on a floodplain or delta. It contains a few fossil plants and some vertebrate fossils. Along with rocks above and below it, the Abo Formation as a whole dips southeastward as the backslope of the Manzano Range, which forms the eastern boundary of the Rio Grande Rift in this area. In some places close to the mountains it is to some extent overridden, along reverse faults, by the much older Precambrian granite that cores the range. The Manzano Range itself is a single east-tilted fault block

An ephemeral stream, flowing only after rain, has developed a plunge pool below a resistant sandstone layer.

National Park Service photo.

about 72 kilometers (45 miles) long. East-dipping sedimentary layers form a series of cuestas along the eastern side of the range.

Many of the blocks used in construction of the village and church are, like those at Gran Quivira, naturally squared by breaking along joints and bedding planes.

Quarai. Constructed in the thirteenth and fourteenth centuries, Quarai also is built of naturally squared blocks of Abo Formation sandstone. Like Abo, this community had a reliable water supply—several seeps unaffected by any but the severest drought. Here as at Abo the red rocks of the Abo Formation dip southeastward off the Manzano Range.

GEOLOGIC HISTORY

Precambrian Era. There are no Precambrian rocks within the monument areas, but those of the Manzano Range, well exposed on the west side of the range, include highly altered sedimentary and igneous rocks as well as relatively unaltered granite. The rocks record repeated folding and faulting, suggesting that several generations of Precambrian mountains formed here, perhaps in response to drifting of a Precambrian continent.

Paleozoic Era. Sedimentary rocks that overlie these Precambrian rocks range in age from Mississippian to Permian. Older Paleozoic seas did not reach this area. In much of the range, relatively soft Pennsylvanian siltstone, sandstone, limestone, and conglomerate lie directly on the Precambrian rocks.

The most striking of the Paleozoic units is the Abo Formation, bright red Permian sandstone, siltstone, and shale exposed in the hills near Abo and Quarai. Formed on a delta or floodplain, the Abo Formation underlies other Permian rocks: the Yeso Formation (a near-shore, shallow-water series of sandstones and siltstones), the Glorietta Sandstone (derived from beach sand), and the San Andres Formation (the marine limestone well exposed around Gran Quivira). These four formations tell us that in Permian time the land was slowly subsiding. Rivers bringing sediments to the region were unable to make up for the sinking of the land, and eventually a shallow sea advanced across the low-lying continental shelf. Late in Permian time, however, the sea retreated westward once more.

Mesozoic Era. Mesozoic rocks are not represented within the three monument units, but exposures along the east side of the Sandia Mountains, north of the Manzano Range, tell us that they did once exist in this part of New Mexico. Triassic and Jurassic units are a colorful series of soft, easily eroded shale and sandstone layers, full of volcanic ash, very like rocks of similar age in the Painted Desert and Petrified Forest of Arizona. Cretaceous units, from the Da-

National Park Service photo.

Fine ripplemarks in Abo Formation sandstone show that it was deposited by flowing water, possibly on a tidal flat.

kota Sandstone (a beach sandstone) upward, show us that the sea returned in Cretaceous time, but from the east rather than from the west.

Cenozoic Era. The withdrawal of the Cretaceous sea signaled the beginning of development of the Rio Grande Rift, as well as the rise of the Rocky Mountains farther north. Subsidence of the long, narrow Rio Grande Rift was coupled with uplift of ranges along its margins, the Manzano Range among them. Its western face broken by the faults that edge the great rift, the Manzano Range hinged upward like a sloping cellar door, its rock strata broken on the west and tilting eastward on the east side of the range. Erosion quickly attacked the rising mountains, washing rock debris into the intervening basins, among them Estancia Basin west of the Manzano Range.

Like many another downfaulted block in this region, the Estancia Basin had no outlet, no through drainage. During the rainy cycles of Ice Age time, fluctuating lakes occupied the depression. As much as 95 meters (300 feet) deep at times, the lakes nevertheless did not overflow the sill at the southeast end of the valley; their waters were therefore in all probability salty. Beaches, bars, and spits can still be distinguished around the margins of the present valley; they show up particularly clearly on aerial photographs. On the lakeshores, ancient campsites show that hunter-gatherers who preceded the pueblo-builders were in this area as early as 12,000 years ago, hunting a now-vanished type of bison, small ancestors of today's horses, mammoths, and perhaps ancestral camels. Possibly these hunter-gatherers contributed to the extinction of some of the large mammals that used to roam this area.

As the climate became drier, the lakes dwindled and disappeared, leaving the dry silt of the lake floor to be whirled skyward by desert winds. As the large mammals disappeared, nomadic hunter-gatherers were supplanted by Plains Indians and the pueblo-builders. As we have seen, Spanish padres arrived in the early years of the seventeenth century. The pueblos and the Spanish churches and convents were abandoned between 1670 and 1690, probably because of prolonged drought, epidemic illness, and raids by warlike Plains Indians.

OTHER READING

Grambling, J. A., and Wells, S. G. (editors), 1982. *Albuquerque Country II*. New Mexico Geological Society Roadguide.

Tonto National Monument

Established: 1907
Size: 4.5 square kilometers (1.75 square miles)
Elevation: 855 meters (2805 feet) at visitor center
Address: P.O.Box 707, Roosevelt, Arizona 85545

STAR FEATURES

• Cliff dwellings occupied in the early fourteenth century, well preserved by overhanging cliffs of Precambrian quartzite.

• Visitor center, introductory slide program, self-guided trail to the lower ruin, guided tour (by prearrangement) to upper ruins.

SETTING THE STAGE

In the rugged setting of Arizona's Mazatzal Mountains—part of a transition region between the Basin and Range deserts to the south and the high plateaus of the canyon country to the north—this monument preserves three groups of 700-year-old cliff dwellings. The small, several-storied villages hug sheltering caves in the Dripping Spring Quartzite, caves that are naturally shaped, the result of gradual weathering of slabby, closely jointed siltstone in the upper part of this formation.

Above the caves and contributing to their preservation are massive cliffs of hard sandstone and siltstone typical of this formation at this locality. On weathered surfaces the rock is tan or light brown; below the surface it is dark with fine grains of carbon and tiny crystals of the mineral pyrite. Much of its surface near and above the caves is hidden by a cementlike coating of conglomerate, held together by the Southwest's common natural cement, caliche. Caliche forms where lime-enriched groundwater, abundant in limestone areas, is drawn to the surface and evaporated, leaving behind small quantities of its mineral burden, which gradually cements the gravel in which it is deposited. Both gravel and caliche are, as we shall see, much younger than sandstone and siltstone of the Dripping Spring Quartzite. Below the caves is a talus slope made up of angular, unsorted rock fragments fallen from the cliffs and steep slopes above.

The Dripping Spring Quartzite is part of a larger rock unit, the Apache Group, which makes up this part of the Mazatzal Mountains. Rock strata below the quartzite (which in this area includes sandstone, claystone, and siltstone) consist of a sequence of dark red, almost maroon siltstone layers, visible near the bottom of Cholla Canyon below the visitor center. Above the quartzite—hence above the ruins—is the Mescal Limestone, here not true limestone but light gray dolomite, a sedimentary rock similar to limestone but with added magnesium carbonate. At

Cliff dwellings of Tonto National Monument occupy caves where strong, normally resistant Precambrian sandstone is weakened by water seeping along slabby bedding surfaces and joints.

the top of the Apache Group and not visible from the national monument is a Precambrian lava flow. All these units play host to numerous dikes and sills, easily spotted on the hillsides because the dark diabase of which they are composed weathers into greenish soil.

The trail from the visitor center to the Lower Ruin zigzags up the talus slope below the ruin. Many of the angular, unsorted rock fragments of the talus display ancient sedimentary features that tell us something about the environment in which the Dripping Spring Quartzite was depos-

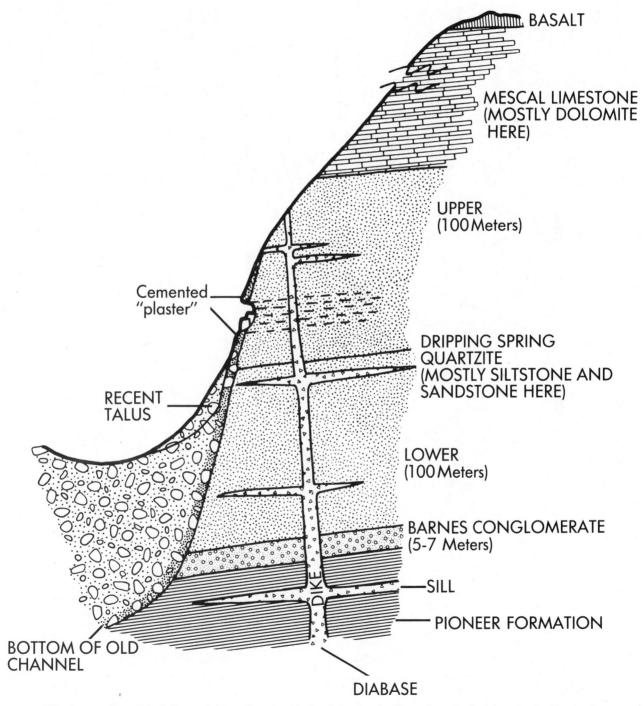

BASALT

MESCAL LIMESTONE (MOSTLY DOLOMITE HERE)

UPPER (100 Meters)

Cemented "plaster"

DRIPPING SPRING QUARTZITE (MOSTLY SILTSTONE AND SANDSTONE HERE)

RECENT TALUS

LOWER (100 Meters)

BARNES CONGLOMERATE (5-7 Meters)

SILL

PIONEER FORMATION

BOTTOM OF OLD CHANNEL

DIABASE

DIKE

Dipping gently eastward, Precambrian sedimentary rocks of the Apache Group form the backdrop for the Tonto ruins.

ited. Some of the ancient rocks display cross-bedding, a patterning diagonal to the originally horizontal rock layers. The cross-bedding is of a type known to be forming today in shallow water, where currents transport sand and deposit it at the edges of small underwater deltas. Billion-year-old ripple marks decorate some rock surfaces. They, too, have modern counterparts, formed by both waves and currents in shallow water. Together with the rock types, they suggest that the muds and sands of this formation were deposited on tidal flats, a view reinforced by the presence of mudcracks, which can form only when and if the original sediments dried out enough to shrink, as they might during ebb

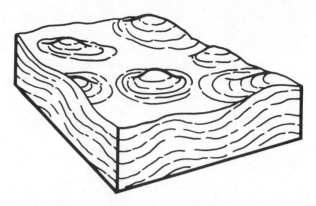

Some of the ancient rocks at Tonto display stromatolites, mounds formed by growing algae. They are among the oldest fossils known.

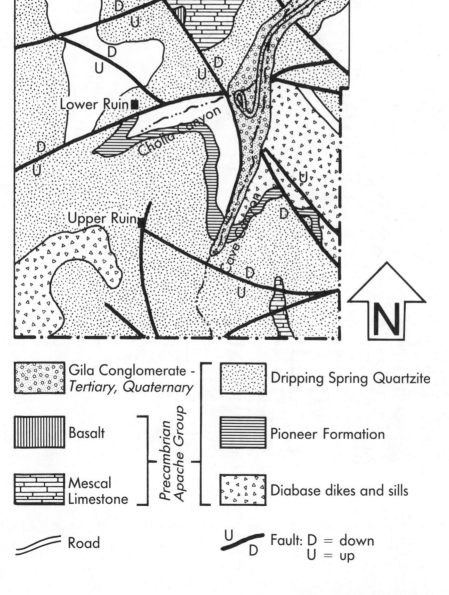

Gila Conglomerate -
Tertiary, Quaternary

Dripping Spring Quartzite

Basalt

Pioneer Formation

Mescal
Limestone

Precambrian Apache Group

Diabase dikes and sills

Road

Fault: D = down
U = up

A geologic map shows the distribution of rock layers in the southeast part of Tonto National Monument.

The cave ceiling at the Lower Ruin shows the slaty or slablike quartzite of the Dripping Springs Formation, cut by vertical joints that may have aided in cave formation.

tides. The mudcracks seen along the trail are really casts of the original mudcracks, preserved in sandstone layers that were deposited on top of the dried-out mud.

The rocks also contain, in places, jagged, toothlike suture lines called stylolites, probably formed by pressure and gradual solution after the sediments had compacted and hardened.

Rock layers here dip gently toward the southwest, away from the cave-dwelling side of the canyon. Although the Lower Ruin and Upper Ruin caves are at different levels in terms of altitude, they are both in the same part of the Dripping Spring Quartzite, in the same zone of slabby siltstone. A fault between the Lower and Upper Ruins offsets this zone, so that one group of dwellings is higher than the other.

At the cave that shelters the Lower Ruin can be seen the slablike nature of this part of the Dripping Spring Quartzite. Note also the zones of closely spaced vertical joints caused by shearing as parts of the rock moved against each other. Both the joints and the cracks between rock slabs contributed to cave formation by allowing groundwater to pass more easily through the rock.

The caves were formed mostly by a process known as spalling, breaking loose of thin crusts of rock, which then tumble onto the floor of the cave or slide down the talus below. Spalling can be caused by a number of factors; here it is due largely to the presence of moisture in the rock. Water dissolves the cementing minerals between grains of sand and silt, weakening the rock. Moreover, water expands when it freezes (as it does here on many winter nights), repeatedly widening joints and other cracks. Here where the siltstone is thinly layered and slabby, and finely sliced in several shear zones, spalling is especially effective. (Small but still-growing stalactites within the caves show that groundwater continues to move through the rock here even today.) It is aided as well by the coating of hardened gravel that covers the cliff face—an unusual natural plaster that holds moisture inside the rock. As we shall see, this coating is part of the Gila Conglomerate, a coarse gravel well cemented with calcium carbonate. It, too, may have helped to localize cave formation: A few small breaches in the natural dam may have been the points at which cave formation began.

GEOLOGIC HISTORY

Precambrian Era. Very old and very young rock—the Precambrian Apache Group and the Cenozoic Gila Conglomerate—tell an interesting, if incomplete, history here. In Precambrian time the area, like all of the Southwest, underwent repeated periods of mountain-building, repeated periods of erosion, and repeated invasions by the sea. About 1.5 to 1 billion years ago, as the sea crept over an eroded, featureless plain, the Apache Group was deposited—many layers of sandstone, siltstone, mudstone, and limestone formed near shore in a gradually deepening sea.

Around 1.2 billion years ago these rocks—by then pretty well consolidated—were fractured and intruded by a network of diabase dikes and sills. There may have been volcanism at that time, with these dikes and sills perhaps serving as feeders for the molten rock. If mountains were formed during this episode, they were later destroyed by the long period of erosion that characterizes the end of Precambrian time.

Paleozoic and Mesozoic Eras. During the next half-billion years this area was almost certainly invaded again by the sea, with the same sedimentary layers deposited as are now exposed in Grand Canyon's walls—layers of limestone, sandstone, and shale—probably piled on top of the Apache Group. At times the region was lifted a little, and received the sands and silts of river floodplains and deltas. At other times, particularly during the Mesozoic Era, it was part of an ancient Sahara, with drifting dunes swept by the wind. Then, toward the end of the Mesozoic Era and lasting into Cenozoic time, this central Arizona region (as well as the Rocky Mountain region well to the north and east of it) was the site of mountain uplift, folding and faulting, and intrusion of molten magma, some of it rich in copper and other minerals. As these high, rugged ranges were once more cut down by erosion, the Paleozoic and Mesozoic sedimentary rocks of this immediate area were destroyed.

Cenozoic Era. Around 15 million years ago, Basin and Range faulting brought more mountain-building, lifting the present central Arizona ranges as well as highlands in adjacent New Mexico. Among the largest and probably the highest mountains in the area, these ranges extended southeast and northwest in a broad, rugged band. As soon as uplift began, erosion attacked—the same old story, everywhere. Streams cut into the uplands and relayed their rock debris down into surrounding lowlands. The coarsest sediments ended their journey fairly close to the mountain margins, while finer material was carried out into the center of the basins—a situation true in all the basins of the

Southwest. In what was to become the Gila River's drainage basin, these sediments, together with associated lake and playa deposits, formed the Gila Conglomerate. In places 5000 meters (15,000 feet) thick, this conglomerate makes up a vast quantity of gravel by any measurement.

By the end of Pleistocene time the downfaulted valley between the central Arizona ranges and high country to the north and east was filled to the brim. Since the terrain when the Gila Conglomerate was deposited was at least as rough and rugged as it is now, the conglomerate in places lay up against cliffs and steep slopes that edged the fault valley. Hardened gradually by minerals carried in groundwater, the margins of the formation became tightly cemented.

The final chapter in the geologic history of Tonto National Monument is one of erosion spurred partly by continuing uplift, partly by downcutting. Gradually the Gila River cut a route through ranges to the southwest, chiseling a deep channel that enabled it to carry its earlier deposits out to the deserts of the Phoenix area and points west. But remnants of the Gila Conglomerate remain, many of them incongruously high on the sides of the ranges that border the Gila's valley. Here in Tonto National Monument only the firmly cemented margins of the formation are left: the plaster that coats the Dripping Spring Formation near the ruins.

Caliche-cemented gravels of the Gila Conglomerate veneer the cliff face near the Upper Ruin.

White Sands National Monument

Established: 1933
Size: 588 square kilometers (227 square miles)
Elevation: 1189 to 1212 meters (3900 to 3980 feet)
Address: Box 458, Alamogordo, New Mexico 88310

STAR FEATURES

• An undulating sea of alabaster sand blown into glistening snow-white dunes.

• An alkali flat and playa "lake" edged with selenite crystals, sources of the gypsum sand.

• Visitor center exhibits, a loop road into the dune area, road and trail leaflets, and guided excursions on the dunes and to the source of the white sand.

See color pages for additional photographs.

SETTING THE STAGE

In sun-soaked Tularosa Valley, between the San Andres and Sacramento mountains, the White Sands extend over an area about 48 kilometers long and 16 kilometers wide (30 × 10 miles). About half the dune area is in White Sands National Monument.

Though most of the world's dunes are made of grains of quartz or tiny rock fragments, the dunes in this national monument are composed almost entirely of gypsum in the form of the mineral selenite. Keys to the accumulation of the

Ripplemarks commonly run up and down dune slopes, a characteristic useful in identifying ancient wind-deposited sandstone.

dune sand are a plentiful source of gypsum, sparse vegetation held in check by the desert climate, and periodic strong winds.

Only winds blowing 7.5 meters per second (about 15 miles per hour) or more are effective in moving grains of sand. Such winds, nearly always blowing from the southwest, occur here in February, March, and April. As we will see, dune contours depend on these one-directional winds.

Four types of dunes have been recognized in the national monument: dome-shaped, barchan, transverse, and parabolic. Dome-shaped dunes occur on the southwest side of the dune field, where the wind is strongest. Barchan and transverse dunes form in the central part of the dune field, where the wind's energy has already decreased. Parabolic dunes develop along the margins of the field where surrounding desert vegetation catches and anchors some of the drifting sand.

As a measure of the quantity of gypsum in these dunes, the part of the dune field *outside* the national monument could furnish nearly four billion tons for commercial use, a 350-year supply for all of the United States. Gypsum is used in plaster, plasterboard, cement, and plaster of Paris casts for fractured limbs. (A compact, fine-grained version of gypsum, used as ornamental stone, is alabaster.)

Where does all this gypsum come from? West and southwest of the dunes, in the lowest part of the Tularosa Valley, is a broad white playa, an

WIND

Dome-shaped dunes are just what the name implies, low circular or oval mounds. They form where winds are strongest.

Barchan dunes are shaped like crescents, with horns pointing away from the wind. The downwind slope, between the horns, is always steeper than the upwind slope.

Long ridges of transverse dunes run across the wind direction, a string of barchan dunes "holding hands." They, too, display gentle windward slopes and steep leeward slip faces. Sand avalanches down the leeward faces, contributing to the gradual downwind march of the dune.

Parabolic dunes are U-shaped, concave on the upwind side. Long arms, partly anchored by developing vegetation, trail back to windward.

Large selenite crystals (shown also in color pages) "grow" in soil near Lake Lucero. Camera (lower right) shows crystal size.

Knowledge of the crossbedding within modern dunes aids in identifying sandstone formed in ancient dunes. This drawing shows the crossbedding inside an 8 meter high (27-foot) barchan dune trenched for geologic studies. Fine laminae that form on the dune's lee face are divided into sets by avalanching and wind erosion. (After E. D. McKee.)

Along the margins of the dune field, plants help to hold the sand in parabolic dunes.

National Park Service photo.

alkali flat that is the bed of ancient Lake Otero. At its southern, deepest end is a smaller region, Lake Lucero, which after heavy rains may hold as much as 30 centimeters (1 foot) of water. Gypsum makes up most of the deposits in the floor of Lake Lucero. This lake and its deposits are the probable source of much of the gypsum in the White Sands dunes. And in the silty lake deposits that fringe the playa, capillary action draws gypsum-saturated groundwater toward the surface, where in today's arid climate it evaporates, leaving gypsum behind as large, clear, daggerlike selenite crystals that seem to grow right out of the soil. Cracked and shattered by repeated exposure to desert temperature extremes, bombarded by windblown sand, the crystals break down into smaller and smaller pieces. As gusts of wind whirl dust-sized particles into the sky, sand-sized fragments are swept northeastward across the flat playa surface, to be added to the White Sands dunes.

Even with constant addition of new gypsum sand from the selenite crystals and playa floor, the dunes do not grow westward. This is because like many dunes they slowly migrate downwind. Sand grains bounce up the long, gentle windward slopes and gather on the dune crests. Then some of the sand avalanches down the steep leeward slopes. Because most strong winds in this area blow from the southwest, the slow march of the dunes is northeastward. Moving in endless procession some of them advance as much as 8 to 10 meters (25 to 33 feet) per year.

The scenic drive beginning at the visitor center will take you into the very heart of the dunes. A guide to points of interest is in the brochure distributed by the Park Service. You will be driving among both transverse and barchan dunes sepa-

Some plants survive even on the flowing gypsum sand, growing upward as their dune increases in height. As the dune moves on, plant roots hold a column of sand in place.

rated by flat, lightly vegetated interdune areas where the gypsum is hard-packed and damp, darker in color than on the dunes. Ripple marks, animal tracks, gypsum pedestals held together by plant roots, and near the road the footprints of visitors mark this sand sea. Evidence of dune movement is nowhere more apparent than along this route, where dunes regularly encroach on the road.

Monthly tours conducted by a national monument interpreter visit Lake Lucero to see the gypsum flats and the remarkable selenite crystals along its shores.

GEOLOGIC HISTORY

Paleozoic and Mesozoic Eras. In Paleozoic time, shallow seas repeatedly swept across this region, which was by and large an area of low relief, without mountains or significant changes in elevation. In the seas were deposited many layers of marine sedimentary rocks: sandstone, shale, and limestone containing marine fossils. About 250 million years ago an arm of a Permian sea, probably bordered by deserts as the Red Sea is today, slowly evaporated, leaving over this region deposits of salt, gypsum, and other evaporite minerals. Many successive floodings and many successive dryings created thick layers of these minerals. The gypsum deposits alone may have been more than 150 meters (500 feet) thick. Finally in Mesozoic time the land rose above sea level, and the evaporite deposits were covered with sand and mud that accumulated in floodplain or delta environments.

Cenozoic Era. About 65 million years ago, following another advance of the sea, the Earth's crust began to move more actively, a reflection of the westward drift of North America as the Atlantic widened. North of here the Rocky Mountains had pushed up. To the west, after general regional uplift, individual blocks of crust broke

The glaring alkali flat west of the White Sands contains evaporite minerals—salt, gypsum, and others—from Pleistocene Lake Otero.

White Sands' gypsum comes from Permian sedimentary rocks that stripe the San Andres Mountains.

apart and by rising here, tilting there, created many isolated ranges. In what is now southern New Mexico a broad anticline bowed upward, arching from (and including) the present San Andres Mountains on the west and the present Sacramento Mountains on the east. In this domed plateau were the ancient floodplain and delta deposits, and below them, not yet exposed at the surface, the even older layers of marine sandstone, shale, salt, and gypsum.

Early in Miocene time, 23 million years ago, the scene changed again. The whole center section of the great arch—a block 48 kilometers wide and 160 kilometers long (30 × 100 miles)—broke away along two prominent fault zones and sank thousands of meters to form the graben of Tularosa Valley. As a direct result of the faulting, Permian rocks containing salt and gypsum were exposed on the cliffs that edged the graben. Rain and melting snow deposited gravel fans below the cliffs, and swept sand and silt into a succession of lakes that occupied the valley floor. (Even today all material eroded from the surrounding mountains accumulates on the floor of the valley, which has no outlet.) From rocks of the bordering cliffs, salt and gypsum were dissolved and flushed into the basin, there to be deposited with the lake sediments. Much of the gypsum and salt found its way into the groundwater, and slowly saturated all the valley sediments.

Today the Permian rocks that contained the gypsum can be seen in the San Andres Mountains to the west and the Sacramento Mountains to the east. Some of the gypsum was deposited in gypsum-saturated Lake Otero, during the last of the wet, cool climate cycles that were southern counterparts of Ice-Age glaciation. When Lake Otero evaporated less than 12,000 years ago, victim of a drying climate, broad white flats remained to show where it once lay. Diminutive Lake Lucero at the south end of the playa is part of nature's continuing process of producing sand for White Sands' dunes.

OTHER READING

Atkinson, Richard, 1977. *White Sands—Wind, Sand, and Time*. Southwest Parks and Monuments Association.

McKee, E. D., and Douglass, J. R., 1971. *Growth and Movement of Dunes at White Sands National Monument, New Mexico*. U.S. Geological Survey Professional Paper 750D, pp. D108–D114.

McKee, E. D., and Moiola, R. J., 1975. *Geometry and Growth of the White Sands Dune Field, New Mexico*. Journal of Research, U.S. Geological Survey, vol. 3, no. 1, pp. 59–66.

Glossary

accretion—the adding of islands or micro-continents to the forward edge of a drifting continent.

alabaster—fine-grained, compact gypsum used as ornamental stone.

algae—a class of aquatic plants, most of them microscopic but also including common seaweeds. Microscopic forms precipitate much of the calcium carbonate that becomes limestone.

alkali—a general term including a mixture of sodium and potassium carbonates, sodium chloride (salt), and other salts, often found in desert playas.

alluvial—deposited by rivers and streams.

alluvial fan—a sloping mass of stream-deposited gravel and sand at the mouth of a steep, narrow canyon, found particularly in desert areas.

anhydrite—calcium sulfate ($CaSO_4$), a form of gypsum that does not contain water.

anticline—a fold that is convex upward.

antiform—an anticline-like structure in igneous rocks, not necessarily produced by folding.

archaeology—the study of prehistoric man's life and culture.

arroyo—a gully or small canyon, dry most of the time.

ash—fine particles of pulverized rock blown from a volcanic vent.

ash flow—volcanic ash that moves down the side of an erupting volcano rather than rising into an ash cloud.

badlands—rough, gullied topography in arid and semiarid regions, eroded by infrequent but heavy rains.

bajada (ba-HA-da)—the broad slope formed by coalescing alluvial fans, often continuous with the pediment surrounding mountain ranges.

balanced rock—a large rock resting more or less precariously on a narrow base.

barchan dune—a crescent-shaped sand dune with arms pointing downwind.

basalt—dark gray to black volcanic rock poor in silica and rich in iron and magnesium minerals.

basin—a downwarped or, in the southwest particularly, downdropped area filled with sediment eroded from surrounding higher areas.

batholith—a large mass of intrusive rock, more than 100 square kilometers (40 square miles) in surface exposure.

bedrock—solid rock exposed at or near the surface.

bentonite—a soft, porous, light-colored clay mineral formed by decomposition of volcanic ash.

biotite—black mica.

bomb, volcanic—a fragment of molten or semi-molten rock thrown from a volcano.

borax—an ore of boron, found in evaporite deposits of some alkaline lakes.

boulder—a rock fragment with a diameter greater than 25 centimeters (10 inches), usually rounded, transported by running water.

brachiopod—a marine shellfish with two bilaterally symmetric shells. Fossil brachiopods are common in Paleozoic rocks.

breakdown—large blocks of rock material fallen from cave walls and ceilings.

breccia (rhymes with "betcha")—rock consisting of coarse, broken rock fragments imbedded in finer material such as volcanic ash.

bryozoan—a group of branching, corallike invertebrates.

calcite—common rock-forming mineral ($CaCO_3$), the principal mineral in limestone, marble, chalk, and travertine.

calcareous—made of or containing calcium carbonate.

calcium carbonate—calcite ($CaCO_3$), major mineral component of limestone and travertine.

caldera—a broad, basin-shaped volcanic depression formed by explosion or collapse of a magma chamber.

caliche (ca-LEE-chee)—calcium carbonate found on or near the surface of the soil in arid and semiarid climates, deposited as calcium carbonate-bearing water is drawn to the surface and evaporated.

caprock—a resistant layer or rock fragment protecting softer rock below.

carapace—a layer of crushed and sheared rock covering a dome of igneous or metamorphic rock in metamorphic core complexes.

carbonic acid—an acid formed of carbon dioxide and water (H_2CO_3).

cave onyx—banded calcite formed in a cave, looking like onyx when polished.

cave pearls—a round ball of calcite formed in subterranean streams.

cave popcorn or **globularite**—a type of cave ornament.

cavern—a large and commonly complex cave.

chert—a hard, dense form of silica that usually occurs as nodules in limestone.

cinder cone—a small, conical volcano built primarily of loose fragments of bubbly volcanic material thrown from a volcanic vent.

cinders—bubbly, popcornlike volcanic material.

climbing dune—a dune formed by wind piling sand against a cliff or steep slope.

cobble—a rounded rock fragment having a diameter of 6.4–25 centimeters (2.5–10 inches).

columnar jointing—a polygonal joint pattern, caused by shrinkage during cooling, that creates vertical columns in lava and welded volcanic tuff.

conduit—the feeder pipe of a volcano.

conglomerate—rock composed of rounded, water-worn fragments of older rock.

continental—sedimentary rocks deposited on land or in lakes, by streams, wind, or ice.

coral—a group of sea-dwelling animals that may deposit calcium carbonate in large reeflike masses.

crater—the funnel-shaped hollow at the summit of a volcano, from which lava and volcanic ash are ejected.

creep—slow downhill movement of soil and rock.

cross-bedding—slanting laminae between main, horizontal layers of sedimentary rocks.

crust—the outermost shell of the Earth, above the mantle and core.

cuesta—a ridge with one long, gentle slope formed by a tilted, resistant rock layer, and one short, steep slope on the cut edges of that and lower layers.

dacite—volcanic rock with a high proportion of quartz and feldspar.

dendritic drainage—a treelike pattern of branching streams and rivulets.

desert pavement—a natural concentration of closely packed pebbles, the result of winnowing by wind action and sheetwash.

desert varnish—a thin, glossy coating of dark brown or blue-black manganese and iron minerals on rocks in desert regions.

detachment—a large mass of rock and soil that has slid downhill along a faultlike sole. Many detachments formed below the surface.

diabase—a dark gray rock common in sills and dikes, the intrusive equivalent of basalt.

dike—a sheetlike intrusion that cuts vertically or nearly vertically across other rock structures. In igneous rocks, dikes are often called **veins**.

diorite—a medium-gray igneous rock, the intrusive equivalent of andesite.

dip—the direction and degree of tilt of sedimentary layers, measured downward from horizontal.

dolomite—a sedimentary rock consisting of calcium and magnesium carbonates.

dome—an anticline in which rocks dip away in all directions. (See also **lava dome**.)

draperies—cave deposits in the form of rippled curtains.

dripstone—travertine deposited by dripping water, as in stalactites and stalagmites.

echinoderm—marine invertebrate animals, including starfish and sea urchins, with radially symmetrical shells.

entrenched—occupying a trench cut by stream or river erosion.

epidote—a yellowish or greenish mineral found mostly in metamorphic rocks.

epoch—a unit of geologic time, subdividing a period.

era—the largest unit of geologic time.

evaporite—minerals left behind by evaporation of sea or lake water; includes salt, gypsum, and anhydrite.

exfoliation—a process in which concentric crusts of rock break away from a rock surface. Also called **spalling**.

extrusive rock—rock formed of magma which flows out on the surface and solidifies there (also called **volcanic rock**).

faceted ridges—ridges or spurs with an inverted V face produced by faulting along a mountain base.

fan—see **alluvial fan**.

fault—a rock fracture along which displacement has occurred.

fault block—a segment of the Earth's crust bounded on two or more sides by faults.

fault scarp—a steep slope or cliff formed by movement along a fault.

fault zone—a zone of numerous small fractures that together make up a fault.

feldspar—a group of common, light-colored, rock-forming minerals containing aluminum oxides and silica. Feldspars constitute 60% of the Earth's crust.

floodplain—relatively horizontal land adjacent to a river channel, with sand and gravel layers deposited by the river during over-the-bank floods.

flowstone—travertine deposited in caves by water trickling across cave walls and floor.

foliated—a stacked-leaves pattern due to alignment of flat mineral crystals or fine, streaky, flaky, or closely spaced joints.

fold—a curve or bend in rock strata.

forereef—the seaward side of a reef, commonly a steep slope faced with reef debris.

formation—a mappable unit of stratified rock.

fossil—remains or traces of a plant or animal preserved in rock; also long-preserved inorganic structures such as fossil ripple marks.

freeze-and-thaw weathering—prying apart of rock by crystal expansion as water freezes repeatedly in rock crevices. Also called **frost wedging**.

fumarole—a vent through which volcanic gases or vapors are emitted.

fusulinid—a group of small, one-celled marine animals of which the shells are shaped like grains of wheat.

gabbro—a dark gray to black, crystalline igneous rock, the intrusive equivalent of basalt.

glacier—a large mass of ice driven by its own weight to move slowly downslope or outward from a center.

globularite—a type of cave ornament that resembles popcorn.

gneiss—banded metamorphic rock formed from granite (which it commonly resembles), sandstone, and other continental rocks.

graben—an area dropped down between two more or less vertical, parallel or nearly parallel faults.

granite—a coarse-grained igneous intrusive rock composed of chunky crystals of quartz and feldspar peppered with dark biotite and/or hornblende. Also, in a broader sense, any light-colored, granular intrusive rock.

gravel—a mixture of more or less rounded pebbles, boulders, and sand, not yet consolidated into rock.

groundwater—subsurface water, as distinct from rivers, streams, seas, and lakes.

group—a major rock unit consisting of two or more formations having certain characteristics in common.

grus—disintegrated granite, its coarse, sandlike texture and angular grains derived directly from the parent rock.

gypsum—a common evaporite mineral, calcium sulphate ($CaSO_4 \cdot H_2O$).

helictite—a curly cave ornament.

hematite—a common dark red iron oxide mineral (Fe_2O_3).

hogback—a steep, sharp-crested ridge with one of two approximately equal slopes formed of a hard caprock, the other of the edges of the caprock and other steeply tilting layers.

hornblende—a black or dark green mineral whose rodlike crystals are common in igneous rocks.

horst—an area lifted between two more or less parallel, near-vertical faults.

hot spring—a spring whose water temperature is higher than body temperature (37° C, or 98.6° F).

hydrostatic pressure—the pressure exerted by water due to the weight of water at higher levels.

Ice Ages—the Pleistocene Epoch. See geologic calendar.

ichthyosaur—an extinct group of swimming reptiles.

igneous rock—any rock formed from molten magma.

intrusive rock—igneous rock created as molten magma intrudes preexisting rocks and cools without reaching the surface.

iridium—a chemical element rare in surface rocks, more common in meteorites and the Earth's mantle.

jasper—a dense, opaque, often colorful variety of chert, a variety of quartz.

joint—a rock fracture along which no significant movement has taken place.

laccolith—a dome due to forceful injection of magma between sedimentary layers, doming upper layers.

lagoon—quiet near-shore water sheltered by a reef or offshore bar.

lava—molten magma that has reached the Earth's surface, or the rock formed when such magma cools.

lava dike—a wall of cooling lava at the edge of a lava flow, to some degree confining the flow.

lava dome—a type of volcano characterized by very thick magma that piles into a rounded dome above its conduit. Also called **volcanic dome**.

lichen—a plant community consisting of a fungus and an alga, appearing as a flat, circular crust on a rock surface.

lily pads—rounded shelves of travertine deposited at the surface of cavern pools or hot springs.

lime—a term commonly, though incorrectly, used for calcium carbonate.

limestone—a sedimentary rock consisting largely of calcium carbonate.

limonite—a yellow-brown iron oxide mineral $(2Fe_2O_3.3H_2O)$.

lithified—turned to stone.

lithosphere—the Earth's crust plus the upper part of the mantle, responding as a unit to forces in the mantle.

low-angle—almost horizontal.

maar crater—a low-relief volcanic crater formed by an explosive steam eruption when hot magma contacts water.

magma—molten rock. When extruded onto the Earth's surface, magma is usually called **lava**.

magma chamber—a reservoir of magma from which volcanic materials are derived, usually only a few kilometers below the surface.

mammoth—an extinct elephantlike mammal.

mantle—the thick, partly molten zone between the Earth's core and crust.

marble—a metamorphic rock derived from limestone.

marine—formed in the sea.

mastodon—an extinct elephantlike mammal.

meander—a loop on the sinuous course of a river.

mesa—a large flat-topped hill with a resistant caprock and steep slopes, larger than a butte but smaller than a plateau.

metamorphic core complex—a type of mountain formed by alteration of a large granite or gneiss dome deep below the surface, probably by low-angle extension faulting.

metamorphic rock—rock derived from pre-existing rocks as they are altered by heat, pressure, and other processes.

metasedimentary rock—sedimentary rock altered by heat, pressure, and other processes but still retaining some sedimentary characteristics.

mica—a group of complex silicate minerals characterized by shiny, closely spaced, parallel layers that can be split apart easily.

microcontinent—an island or submerged mass of continental type rock, a fragment of a continent.

mineral—a naturally occurring inorganic substance with a characteristic chemical composition and frequently with typical color, texture, and crystal form.

mollusk—an animal group including clams and snails as well as octopus and squid.

moonstone—a feldspar mineral valued for its silky, pearly translucence.

mud cracks—shrinkage cracks in drying mud.

normal fault—a fault in which the hanging (upper) wall moves downward relative to the footwall.

obsidian—black volcanic glass.

olivine—a green mineral common in mantle-derived basalt and gabbro.

opal—a silica mineral containing up to 20% water, often with an iridescent play of color.

outcrop—bedrock that appears at the surface.

overthrust—a low-angle fault in which one part of the crust slides over another, placing older rock on top of younger; also used for the oversliding block.

paleomagnetic dating—dating of rocks by comparison with known patterns of reversal of the Earth's magnetic field, when the north and south poles switch positive and negative magnetic charges.

Pangaea—a supercontinent composed of all the present continents joined together.

parabolic dune—a sand dune with a convex form, the arms trailing to windward and often held in place by vegetation.

pebble—a rock fragment, commonly rounded, 0.4 to 6.4 centimeters (0.2 to 2.5 inches) in diameter.

pediment—a gently inclined erosion surface carved in bedrock at the base of a mountain range.

pegmatite—exceptionally coarse-grained igneous rock found as dikes or veins in large igneous intrusions.

period—a subdivision of geologic time shorter than an era, longer than an epoch.

phyllite—shiny metamorphic rock having abundant mica crystals, intermediate between slate and mica schist.

plagioclase—a mineral of the feldspar group.

plate—a block of the Earth's crust, separated from other blocks by mid-ocean ridges, trenches, and/or collision zones.

plateau—a flat-topped tableland more extensive than a mesa.

Plate Tectonic Theory—a theory that explains why sea floors spread and continents move apart as new crust is created at mid-ocean ridges.

playa—a flat-floored, vegetation-free lakebed that dries up quickly after rains, characteristic of desert basins with no external drainage.

playa lake—a shallow intermittent lake occupying a playa.

pluvial—relating to rain, particularly southwestern rainy cycles that accompanied glacial advances farther north.

pluton—a single igneous intrusion.

porphyry—igneous rock that contains conspicuous large crystals (**phenocrysts**) in a fine-grained matrix.

pothole—a small but deep circular depression excavated by the grinding action of pebbles, cobbles, and sand swirled by running water.

pressure-release joint—a joint concentric with the surface of once-buried rock, forming by release of pressure as overburden is washed away.

pterodactyl—an extinct flying reptile.

pumice—light-colored, frothy volcanic rock, often light enough to float on water.

pyrite—a common brass-yellow iron mineral (FeS_2) also known as "fool's gold."

quartz—crystalline silica (SiO_2), a common rock-forming mineral.

quartzite—sandstone consisting chiefly of quartz grains welded so firmly that it breaks through rather than around the grains.

quartz monzonite—a granitic rock containing a large proportion of certain potassium and sodium feldspar minerals.

radiometric dating—dating of rocks by analysis of their radioactive minerals and substances (daughter products) formed by decay of those minerals.

reef—a ridge bordering a sea or lake, formed of accumulated remains of shell-secreting plants and animals.

reverse fault—a fault in which the hanging (overhanging) wall moves upward relative to the footwall.

rhyodacite—extrusive igneous rock intermediate between dacite and rhyolite.

rhyolite—light gray volcanic rock with large quartz and feldspar crystals in a finer groundmass, the fine-grained extrusive equivalent of granite.

rift valley—a graben of considerable size, bordered by faults that reach through the Earth's crust to the mantle, as shown by the basalt composition of associated volcanoes.

ripple mark—a pattern of small ridges and hollows formed by water or wind flowing over a sand or silt surface.

rockslide—a landslide involving a large proportion of rock.

salt pan—an alkali flat or playa surfaced with salt and other evaporite minerals.

schist—crystalline metamorphic rock which splits easily along parallel planes, commonly formed from fine-grained sedimentary rock.

scoria—very bubbly volcanic rock, darker and heavier than pumice and with larger bubble holes.

sea-floor spreading—movement of oceanic crust

away from mid-ocean ridges and creation of new oceanic crust at the ridges.

sedimentary rock—rock composed of particles of other rock transported and deposited by water, wind, or ice.

sediment—fragmented rock, as well as shells and other animal and plant material, deposited by wind, water, or ice.

seismic—having to do with earthquakes or earth tremors.

selenite—a clear variety of gypsum commonly found in veins.

shale—fine-grained mudstone or claystone that splits easily along bedding planes.

shield volcano—a broadly dome-shaped volcano formed by moderately fluid lava.

shrinkage cracks—cracks formed when mud dries or magma cools.

silica—a hard, resistant mineral (SiO_2) which in its crystal form is quartz. It also occurs as opal, chalcedony, siliceous sinter, chert, and flint.

sill—a flat igneous intrusion that has pushed between layers of stratified rock.

sinkhole—a more or less circular hole created when part of the roof of a near-surface cave collapses.

slickenside—a shiny, grooved surface created as rock moves against rock, often serving to identify faults.

slump—a landslide in which rock and earth slide as a single mass along a curved slip surface.

soda straw stalactite—a delicate, hollow stalactite that forms as water droplets seep from small rock pores.

soil creep—gradual downhill movement of soil and loose rock.

spheroidal weathering—weathering of joint-edged blocks to a more or less spherical shape.

spalling—breaking away of thin surface layers of rock.

stalactite—a cylindrical or conical cave ornament hanging from a cave ceiling.

stalagmite—a cylindrical or columnar cave ornament projecting upward from the floor of a cave.

stock—an intrusion smaller than a batholith, unlike a laccolith in having no known base.

strata—layers of sedimentary (and sometimes volcanic) rocks.

stratified—layered.

stratigraphic—pertaining to the study of strata (and therefore of sedimentary rocks).

stratovolcano—a volcanic mountain built of alternating layers of lava, breccia, and volcanic ash.

stromatolite—a rounded or irregular, layered limestone mass formed by algae.

structure—pertaining to the form and arrangement of rocks, in particular to faults and folds.

stylolite—a jagged, toothlike suture between layers of sedimentary rock.

subduction—the downward plunge of an oceanic plate below a continental plate.

syncline—a fold that is convex downward.

talus—a mass of large rock fragments lying at the base of a cliff or steep slope from which they have fallen.

tapir—a mammal having a heavy body, short legs, and a flexible snout.

tent rocks—cone-shaped rocks, eroded usually in volcanic ash (tuff) and resulting from cementing of the ash near post-eruption steam vents.

thermal gradient—the rate of temperature increase below the Earth's surface.

thrust fault—a low-angle fault on which older rocks slide over younger ones.

titanothere—an extinct mammal of Tertiary time, resembling a rhinoceros.

transverse dune—a ridgelike dune at right angles to the wind direction.

travertine—a hard, dense limestone deposited by lime-laden water of streams, hot springs, and caves.

trench—a seafloor depression marking the zone where oceanic crust dives beneath continental crust.

tuff—rock formed from volcanic ash.

unconformity—a substantial break or gap in the geologic record, caused by an interruption of deposition, by uplift and erosion, or by igneous activity.

valley fill—sand, gravel, and other rock material filling a valley.

vein—a thin, sheetlike intrusion into a crevice, often with associated mineral deposits.

vent—any opening through which volcanic material is ejected.

vesicle—a small hole formed by entrapment of a gas bubble in cooling lava.

viscous—thick and sticky, not flowing easily.

volcanic ash—fine, airborne material ejected by a volcano.

volcanic dome—see **lava dome**.

volcanic hailstones—small spheres of volcanic ash created by mid-air accretion of ash particles.

volcanic neck or plug—solidified lava that cooled within a volcanic conduit, generally more resistant than surrounding rock.

volcanic rock—rock formed from magma exploded from or flowing out on the Earth's surface.

wash—a southwestern term for a normally dry streambed.

water table—the upper surface of groundwater, below which rock is saturated with water.

weathering—changes in rock due to exposure to the atmosphere.

welded tuff—rock formed of volcanic ash fused by its own heat, the heat of volcanic gases ejected with it, and the weight of overlying material.

wind gap—a now-dry notch in a mountain ridge, commonly cut and then abandoned by a stream.

Index

Page numbers in **boldface** indicate major discussions. Page numbers preceded by a "C" refer to the color section.